REVISE EDEXCEL GCSE (9-1)
Mathematics
Higher

PRACTICE PAPERS Plus⁺

Authors: Jean Linsky, Navtej Marwaha and Harry Smith

These Practice Papers are designed to complement your revision and to help prepare you for the exams. They do not include all the content and skills needed for the complete course and have been written to help you practise what you have learned. They may not be representative of a real exam paper. Remember that the official Pearson specification and associated assessment guidance materials are the only authoritative source of information and should always be referred to for definitive guidance.

For further information, go to www.edexcel.com/gcsemathssupport

ALWAYS LEARNING

PEARSON

Contents

Set A example papers

Set B example papers

Using this book

This book has been created to help you prepare for your exam by familiarising yourself with the approach of the papers and the exam-style questions. Unlike the exam, however, each question has targeted hints, guidance and support in the margin to help you understand how to tackle them.

All questions also have fully worked solutions shown in the back of the book for you to refer to. In addition, some questions have videos explaining the working step-by-step. Look out for the QR codes in green boxes. To watch these videos, scan the QR codes with your mobile phone or tablet using a QR reader.

You may want to work through the papers at your own pace, to re-inforce your knowledge of the topics and practise the skills you have gained throughout your course. Alternatively, you might want to practise completing a paper as if in an exam. If you do this, bear these points in mind:

- Use black ink or ball-point pen.
- Answer all questions.
- Answer the questions in the spaces provided – there may be more space than you need.
- In a real exam, **you must show all your working out**.
- For each paper, check whether you can use a calculator or not. This is stated at the start of each paper. You **cannot** use a calculator for Paper 1.
- If your calculator does not have a π button, take the value of π to be 3.142 unless the question instructs otherwise.
- Diagrams are **not** accurately drawn, unless otherwise indicated in the question.
- The total number of marks available for each paper is 80 marks.
- You have 1 hour 30 minutes to complete each paper.
- The marks for each question are shown in brackets. Use this as a guide as to how much time to spend on each question.

Paper 1: Non-calculator
Time allowed: 1 hour 30 minutes

1   Work out 3.25×0.46

....
(Total for Question 1 is 3 marks)

NUMBER

Revision Guide
Pages 6, 7

Hint

Estimate the answer before you start. You can use your answer to check that you have the right number of decimal places. To estimate, round both numbers to 1 significant figure then multiply:

$3.25 \rightarrow 3$ and $0.46 \rightarrow 0.5$ so your estimate will be 3×0.5

LEARN IT!

To multiply decimal numbers without a calculator:

- ignore the decimal points and just multiply the numbers
- count the number of decimal places in the calculation
- put this number of decimal places (including zeroes) in the answer.

Turn to page 124 for complete worked solutions to the questions on this page.

PROBABILITY & STATISTICS

Revision Guide Page 125

Hint

If you are working out values for a Venn diagram, always start at the centre and work out. You know that 6 children sing in the choir **and** play in the band so you can write '6' in the centre of your Venn diagram.

Hint

You can check a Venn diagram by adding up all the numbers. The total for this diagram should be 30

Hint

For part **(b)**, you are interested in this section of the Venn diagram:

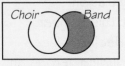

2 There are 30 children in a class.

21 of the children sing in the choir.
10 of the children play in the band.
6 of the children sing in the choir **and** play in the band.

(a) Complete the Venn diagram to show this information.

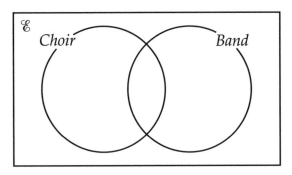

(3)

One of the children from the class is chosen at random.

(b) Work out the probability that this child plays in the band, but does **not** sing in the choir.

....

(2)

(Total for Question 2 is 5 marks)

Turn to page 124 for complete worked solutions to the questions on this page.

3   The diagram shows the area of each of three faces of a cuboid.

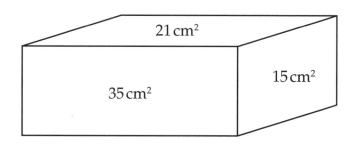

The length of each edge of the cuboid is a whole number of centimetres.

Work out the volume of the cuboid.

Problem solved!

The dimensions of the front face are whole numbers with a product of 35. The only possibilities are 1×35 or 5×7. Find factor pairs of 21, 35 and 15 to work out the dimensions of the cuboid.

Watch out!

The diagram is not accurate, so you can't measure. Just because one side **looks** longer than another, it doesn't mean that it is.

Hint

Once you have worked out the dimensions of the cuboid, write them on the diagram.

Explore

The length of each edge is the HCF of the areas of the faces that meet at that edge.

 Scan this QR code for a video of this question being solved!

 .... cm^3

(Total for Question 3 is 4 marks)

Turn to page 124 for complete worked solutions to the questions on this page.

PROBABILITY
& STATISTICS

Revision Guide
Pages 118, 122

Hint

For a cumulative frequency diagram, you plot the points at the **top** of each class interval. So the first two points you need to plot are (20, 0) and (40, 7).

Watch out!

Join your points with a smooth curve, not straight lines.

Hint

There are 74 data values. Read across from $\frac{74}{4} = 18.5$ for the lower quartile, $\frac{74}{2} = 37$ for the median and $\frac{3 \times 74}{4} = 55.5$ for the upper quartile.

4   When a person exercises, their pulse rate increases.

The time it takes for their pulse rate to return to normal after exercise is called the recovery time.

A group of people did some exercise.

The table below shows some information about their recovery times.

Recovery time (t seconds)	Cumulative frequency
$0 < t \leqslant 20$	0
$0 < t \leqslant 40$	7
$0 < t \leqslant 60$	16
$0 < t \leqslant 80$	34
$0 < t \leqslant 90$	47
$0 < t \leqslant 100$	59
$0 < t \leqslant 120$	68
$0 < t \leqslant 140$	74

(a) On the grid below, draw a cumulative frequency graph for this information.   **(2)**

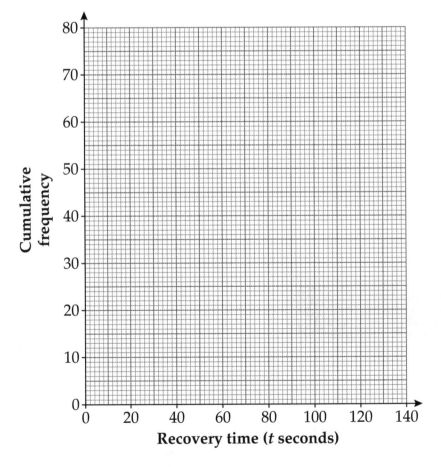

Turn to page 124 for complete worked solutions to the questions on this page.

A different group of people did the same exercise.

Their recovery times had a median of 61 seconds and an interquartile range of 22 seconds.

(b) Compare the recovery times of these two groups of people.

Problem solved!

Always give **evidence** when you are comparing two distributions. For a cumulative frequency diagram, compare the **median** and the **interquartile range**. Make sure your statements refer to the **context** of the question. You need to talk about **recovery times**.

Watch out!

Don't waste time writing long-winded answers. Compare values for each distribution, then write a short conclusion.

Explore

Many distributions follow a 'bell curve', with more data values in the middle. This is why cumulative frequency graphs are often steeper in the middle and shallower at each end.

(5)

(Total for Question 4 is 7 marks)

Turn to page 125 for complete worked solutions to the questions on this page.

Revision Guide
Pages 19, 80

Watch out!

Unless it says so in the question, diagrams in your exam are **not accurate**. You can't measure any lengths for this question – you need to use algebra.

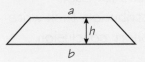

Area of trapezium
= $\frac{1}{2}(a + b)h$

Problem solved!

Don't panic if lengths or angles are given as **expressions** rather than numbers. You can substitute an expression into a formula in exactly the same way as a number. Use the fact that the two areas are equal to write an equation, then solve it to find x.

Watch out!

Make sure your final answer is the **length** of the **rectangular** desk.

---

5 A company makes two different desks.

The top of one desk is in the shape of a trapezium.
The top of the other desk is in the shape of a rectangle.

The diagram shows the tops of the two desks.

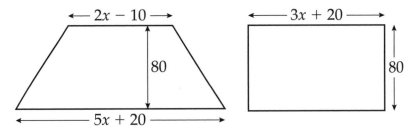

All measurements are in centimetres.

The tops of the two desks have the same area.

Work out the length, in centimetres, of the rectangular desk.
You must show all your working.

.... cm

(Total for Question 5 is 5 marks)

Turn to page 125 for complete worked solutions to the questions on this page.

6   Martin's house has a meter to measure the amount of water he uses. Martin pays on Tariff A for the number of water units he uses.

The graph on the following page can be used to find out how much he pays.

(a) (i) Find the gradient of this line.

....
(2)

Martin reduces the amount of water he uses by 15 units.

   (ii) How much money does he save?

£....................................
(1)

√xy²   ALGEBRA

Revision Guide
Page 30

Hint

Draw a triangle to find the gradient of a graph:

$$\text{Gradient} = \frac{\text{Distance up}}{\text{Distance across}}$$

The larger your triangle, the more accurate your calculation will be.

Hint

The gradient tells you how much the cost increases or decreases for each unit of water. You can work out the answer to part **(a)(ii)** by multiplying your gradient by 15

Turn to page 125 for complete worked solutions to the questions on this page.

Hint

To draw the graph
for Tariff B, plot the
points from the table
and then join them up
using a ruler. The point
where the graphs cross
represents the point
where both tariffs cost
the same. Read down
from this point to the
horizontal axis.

Watch out!

Always use the scale to
work out the distance
up and the distance
across – don't just
count grid squares on
the graph.

Hint

Always read graphs
accurate to the
nearest small grid
square. You need to
use a sharp pencil to
get accurate readings.

Instead of Tariff A, Martin could pay for his water on Tariff B.

The table shows how much Martin would pay for his water on Tariff B.

Number of water units used	0	20	40	60	80	100
Cost in £	12	18	24	30	36	42

(b) (i) On the grid, draw a line for Tariff B.

(2)

(ii) Write down the number of water units used when the cost
of Tariff A is the same as the cost of Tariff B.

.... units

(1)

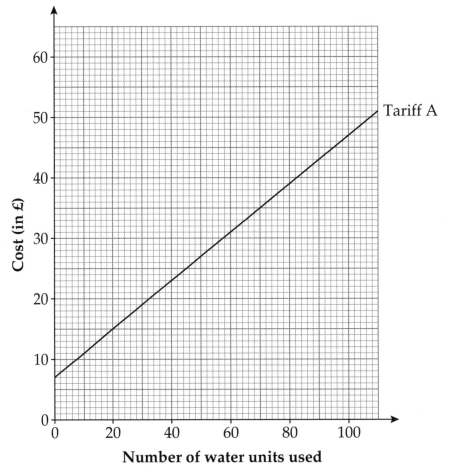

(Total for Question 6 is 6 marks)

Turn to page 125 for complete worked solutions to the questions on this page.

7   ABCD and PQRS are two rectangles.

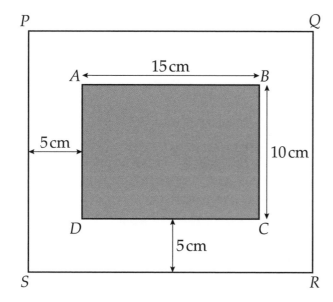

Rectangle ABCD is 15 cm by 10 cm.
There is a space 5 cm wide between rectangle ABCD and rectangle PQRS.

Are rectangle ABCD and rectangle PQRS mathematically similar?
You must show how you got your answer.

(Total for Question 7 is 3 marks)

GEOMETRY AND MEASURES

 Revision Guide Page 97

LEARN IT!

If two shapes are **similar** then corresponding sides will be in the same ratio.

Problem solved!

You need to show enough working to explain your answer. Work out the dimensions of rectangle PQRS, then check for similarity. You could compare ratios between the shapes $\left(\text{e.g. } \dfrac{AB}{PQ} \text{ and } \dfrac{AD}{PS}\right)$ or ratios on each shape $\left(\text{e.g. } \dfrac{PS}{PQ} \text{ and } \dfrac{AD}{AB}\right)$.

Hint

Similar shapes are enlargements of each other.

Turn to page 126 for complete worked solutions to the questions on this page.

9

NUMBER

 Revision Guide
Pages 2, 3

Hint

Make sure you are confident manipulating powers without a calculator.

LEARN IT!

You need to know the squares up to 15^2 and the cubes of 1, 2, 3, 4, 5 and 10. You also need to know the corresponding square roots and cube roots.

Hint

$27^{\frac{2}{3}} = \left(\sqrt[3]{27}\right)^2$

---

8   (a)   Write down the value of $10^{-1} \times 5^0$

....

(2)

(b)   Find the value of $27^{\frac{2}{3}}$

....

(2)

(Total for Question 8 is 4 marks)

Turn to page 126 for complete worked solutions to the questions on this page.

9 Simplify fully $\dfrac{3x^2 - 6x}{x^2 + 2x - 8}$

 **ALGEBRA**

Revision Guide
Pages 18, 47

Hint

To simplify a single
algebraic fraction
you usually have to
factorise the top, the
bottom or both.

Explore

When you simplify
an algebraic fraction
by cancelling a linear
factor, you should
really **exclude** any
values that make that
factor equal to 0.
$\dfrac{3(x - 5)}{x(x - 5)} = \dfrac{3}{x}$, but only
for $x \neq 5$. When $x = 5$
the left-hand side is
not defined, because
you can't divide by 0.

You don't have to
worry about this in
your GCSE exam.

....
(Total for Question 9 is 3 marks)

Turn to page 126 for complete worked solutions to the questions on this page.

RATIO AND PROPORTION

 Revision Guide
Page 65

LEARN IT!

Sketch the formula triangle for speed on your working:

Watch out!

Be careful! Five hours is the time of the **whole journey**. You need to find the time for Appleton to Brockley before you can work out the distance of that part of the journey.

Problem solved!

Think about what information you need to answer the question. You need Harry's total distance travelled, and total time taken.

Watch out!

Check that your answer makes sense as an average driving speed.

---

10 Harry travels from Appleton to Brockley at an average speed of 50 mph. He then travels from Brockley to Cantham at an average speed of 70 mph.

Harry takes a total time of 5 hours to travel from Appleton to Cantham.
The distance from Brockley to Cantham is 210 miles.

Calculate Harry's average speed for the total distance travelled from Appleton to Cantham.

.... mph
(Total for Question 10 is 4 marks)

   Turn to page 126 for complete worked solutions to the questions on this page.

11

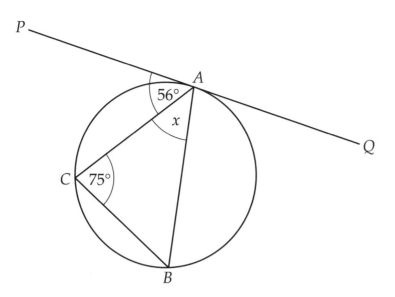

GEOMETRY
AND MEASURES

Revision Guide
Page 105

A, *B* and *C* are points on the circumference of a circle.
The straight line *PAQ* is a tangent to the circle.
Angle *PAC* = 56°
Angle *ACB* = 75°

Work out the size of the angle marked *x*.
Give reasons for each stage of your working.

LEARN IT!

If you see a **tangent** and a **chord** in a circle question, you might be able to use the **alternate segment theorem**.

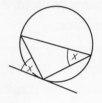

Hint

Write any angles you work out on your diagram as you go. Remember, you still need to write down reasons for each step of your working.

Scan this QR code for a video of this question being solved!

..................................°

(Total for Question 11 is 3 marks)

Turn to page 127 for complete worked solutions to the questions on this page.

13

Revision Guide
Page 8

Hint

You need to work out
$(2.4 \times 10^{10}) \div (6 \times 10^{-2})$

Hint

To divide standard form numbers **without a calculator**:

- divide the number parts
- divide the powers of 10
- rewrite in standard form if necessary.

Watch out!

Remember the third step in the hint above. A number is only in standard form if the number part is **greater than or equal to 1 and less than 10**

---

12 The number 2.4×10^{10} is bigger than the number 6×10^{-2}

How many times bigger?
Give your answer in standard form.

Scan this QR code for a video of this question being solved!

....

(Total for Question 12 is 2 marks)

Turn to page 127 for complete worked solutions to the questions on this page.

13

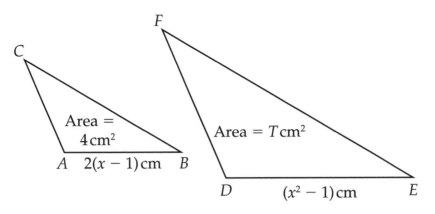

Triangles ABC and DEF are mathematically similar.

The base, AB, of triangle ABC has length $2(x - 1)$ cm.
The base, DE, of triangle DEF has length $(x^2 - 1)$ cm.

The area of triangle ABC is 4 cm^2.
The area of triangle DEF is T cm^2.

Prove that $T = x^2 + 2x + 1$

(Total for Question 13 is 4 marks)

$\sqrt{\mathrm{xy}^2}$ **ALGEBRA**

GEOMETRY AND MEASURES

 Revision Guide Pages 16, 98

LEARN IT!

For similar shapes, if length is multiplied by k, then area is multiplied by k^2.

Hint

Work out an expression for the length multiplier, k. Then square it, and multiply by the area of the smaller triangle.

Problem solved!

If you can't see how to start with this question, try it with numbers instead. Can you show that $T = 25$ in this example:

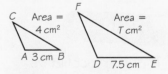

Now go back to the algebra.

Hint

It's always a good idea to write down what you have proved. The last line of your working could be, "So $T = x^2 + 2x + 1$".

Turn to page 127 for complete worked solutions to the questions on this page.

14 Here is the graph of $y = f(x)$.

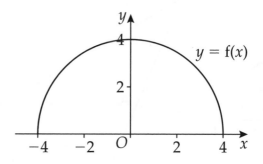

(a) Write down the coordinates of the point P where the graph of $y = f(x) - 3$ meets the y-axis.

(..............., ...)
(1)

The graph of $f(x + 3)$ meets the negative x-axis at the point A.

(b) Work out the area of triangle AOP.

.... square units
(3)

(Total for Question 14 is 4 marks)

    Turn to page 127 for complete worked solutions to the questions on this page.

15 $(2x^{\frac{3}{2}}y^{-1})^n = A\,x^6\,y^B$

Work out the value of n, the value of A and the value of B.

NUMBER

ALGEBRA

Revision Guide
Pages 2, 16

LEARN IT!

Use $(ab)^n = a^n b^n$ to simplify the left-hand side.

Hint

The order of the variables in the answer line might give you a clue about the best order to work them out.

Problem solved!

You can **equate powers** of x and y in this question. Simplify the left-hand side, then set the power of x equal to 6. Solve the resulting equation to find n.

Hint

The question asks for the **values** of n, A and B, so all your answers will be numbers.

$n =$ ....

$A =$ ....

$B =$ ....

(Total for Question 15 is 3 marks)

Watch out!

Don't leave your value of A as a power of 2. Work it out as an ordinary number.

Turn to page 128 for complete worked solutions to the questions on this page.

RATIO AND PROPORTION

 Revision Guide
Pages 69, 70

LEARN IT!

If R is inversely proportional to the square of r, the proportionality formula will look like:

$$R = \frac{k}{r^2}$$

Hint

Find the value of k to write a formula for R in terms of r. Use this to find R when $r = 5$ and when $r = 10$

Watch out!

You need to calculate two resistances, then show that the **difference** is 13.5 ohms.

Explore

The resistance of a wire is inversely proportional to its cross-sectional area. If the cross-section is a circle, then the resistance is inversely proportional to the **square** of the radius.

---

16 The resistance, R ohms, of a particular cable is inversely proportional to the square of its radius, r mm.

When the radius is 3 mm the resistance is 50 ohms.

Cable A has a radius of 5 mm.
Cable B has a radius of 10 mm.

Show that the difference in the resistance of two cables is 13.5 ohms.

Scan this QR code for a video of this question being solved!

(Total for Question 16 is 3 marks)

Turn to page 128 for complete worked solutions to the questions on this page.

17 The diagram shows an equilateral triangle *ABC*.

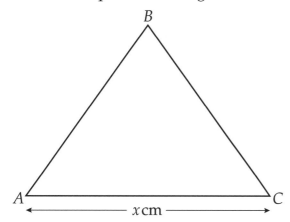

Show that the area of the triangle can be written as $\dfrac{x^2\sqrt{3}}{4}$

(Total for Question 17 is 2 marks)

Turn to page 128 for complete worked solutions to the questions on this page.

GEOMETRY AND MEASURES

Revision Guide
Pages 79, 101

Hint

This is an equilateral
triangle, so it has three
equal angles.

LEARN IT!

Area $= \dfrac{1}{2}\, ab \sin C$

Watch out!

You need to know the
exact values of sin,
cos and tan of 30°,
45° and 60° **without a
calculator.**

Explore

You can use this right-
angled triangle to find
sin, cos and tan of 30°
and 60° without your
calculator.

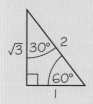

GEOMETRY
AND MEASURES

 Revision Guide
Pages 106, 107

Hint

Write vectors between two points with an arrow. $\overrightarrow{OA}$ is the **vector** from O to A, whereas OA is the **line segment** between O and A.

Hint

For part **(a)**, trace a path from A to B. If you go **backwards** along a vector you have to **subtract** that vector.

Hint

Don't write **a** and **b** in bold in your answer! Just use normal neat letters.

Problem solved!

For part **(b)**, find an expression for $\overrightarrow{CD}$ in terms of **a** and **b**.
If you can show that $\overrightarrow{CD} = k\overrightarrow{OA}$ then the lines CD and OA are **parallel**, and the length of CD is k times the length of OA.

Watch out!

The final line of your answer should be about **line segments**, not vectors.

18

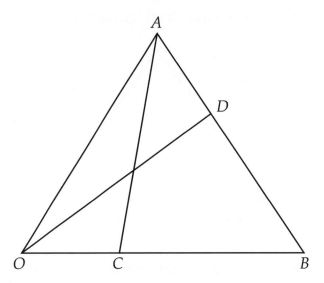

OAB is a triangle.

The point *D* divides the line *AB* in the ratio 1 : 2
The point *C* divides the line *OB* in the ratio 1 : 2

$\overrightarrow{OA} = 6\mathbf{a}$
$\overrightarrow{OB} = 6\mathbf{b}$

(a) Write down $\overrightarrow{AB}$ in terms of **a** and **b**.

....
(1)

(b) Show that:

CD is a parallel to OA **and** the length of CD is $\frac{2}{3}$ the length of OA.

(5)
(Total for Question 18 is 6 marks)

Turn to page 128 for complete worked solutions to the questions on this page.

19 A function is defined by $f(x) = \dfrac{x-1}{x+2}$, $x \in R$, $x \neq -2$

(a) Find $f^{-1}(x)$.

$$f^{-1}(x) = \text{.................................}$$

(3)

(b) Show that $f^{-1}(x) = -2$ has no solutions.

(2)

(Total for Question 19 is 5 marks)

Turn to page 129 for complete worked solutions to the questions on this page.

 √xy² ALGEBRA

Revision Guide
Pages 46, 51

Hint

Write the function in the form $y = \ldots$ then rearrange to make x the subject.

Hint

x appears twice in the equation. You will need to group the x terms on one side then factorise to get x on its own.

Problem solved!

For part **(b)** you should attempt to solve the equation. Show all your working, and when you reach a point where you can't continue, write a short conclusion explaining why there are no solutions.

Explore

The equation $x = x + 1$ has no solutions. If you subtracted x from both sides you would have $0 = 1$ which is impossible.

 Revision Guide
Pages 12, 17

Watch out!

You need to be really careful with brackets and signs (+ or −) in this question. You might find it easier to simplify $(1 + \sqrt{3})^2$ before substituting.

Hint

$(1 + \sqrt{3})^2$

$= (1 + \sqrt{3})(1 + \sqrt{3})$

$= 1 + 2\sqrt{3} + 3$

$= 4 + 2\sqrt{3}$

Hint

Remember that a and b can be positive or negative.

---

20 Given that $x = 1 + \sqrt{3}$, work out the exact value of $\dfrac{11x^2}{3 - 2x}$

Express your answer in the form $a + b\sqrt{3}$ where a and b are integers.

....

(Total for Question 20 is 4 marks)

TOTAL FOR PAPER IS 80 MARKS

22

Turn to page 129 for complete worked solutions to the questions on this page.

Paper 2: Calculator
Time allowed: 1 hour 30 minutes

1   The Lowest Common Multiple (LCM) of three numbers is 30
Two of the numbers are 2 and 5

What could be the third number?

....

(Total for Question 1 is 2 marks)

Revision Guide
Page 1

LEARN IT!

The LCM of three numbers is the smallest number that is a multiple of all three numbers.

Problem solved!

There might be more than one possible answer. You only need to give one possibility.

Turn to page 130 for complete worked solutions to the questions on this page.

23

RATIO AND PROPORTION

Revision Guide
Page 60

Watch out!

You have to read this question really carefully to find the important information. You could try underlining important facts.

LEARN IT!

Add together the numbers in a ratio to work out the **total number of parts**. Divide the total amount by this to work out what **each part is worth**.

Problem solved!

Sometimes you can't work out your complete strategy at the start. See what you **can** work out easily – you could start by working out how much orange **drink** Lily needs in **total**. Or you could start by working out how much orange **squash** is needed for each cup.

Watch out!

Lily needs to buy a **whole number** of bottles of squash.

---

2 140 children will be at a school sports day.
Lily is going to give a cup of orange drink to each of the 140 children.
She is going to put 200 millilitres of orange drink in each cup.

The orange drink is made from orange squash and water.
The orange squash and water are mixed in the ratio 1 : 9 by volume.

Orange squash is sold in bottles containing 750 millilitres.

Work out how many bottles of orange squash Lily needs to buy.
You must show all your working.

....
(Total for Question 2 is 4 marks)

Turn to page 130 for complete worked solutions to the questions on this page.

3 Henri and Ray buy some flowers for their mother.

They buy:

> 2 bunches of roses and 3 bunches of tulips for £10
> 1 bunch of roses and 4 bunches of tulips for £9.50.

(a) Work out the cost of one bunch of tulips.

£ ....

(4)

Henri is 16 years old and Ray is 2 years younger than Henri.
They share the total cost of £19.50 in the ratio of their ages.

(b) Work out how much Henri pays and how much Ray pays.

Henri £ ....

Ray £ ....

(3)

(Total for Question 3 is 7 marks)

 ALGEBRA

RATIO AND PROPORTION

Revision Guide
Pages 34, 60

Problem solved!

For part **(a)** you will need to form two simultaneous equations and then solve them. Use *r* to represent the cost of one bunch of roses, and *t* to represent the cost of one bunch of tulips.

Hint

If you can't do the first part of a question you might still be able to do the second part. The first part needs algebra, but the second part is a ratio question.

Hint

Simplify the ratio before doing any calculations.

LEARN IT!

To divide a quantity in a given ratio:

- add together the parts in the ratio
- divide the amount by the total
- multiply the answer by each part.

Turn to page 130 for complete worked solutions to the questions on this page.

GEOMETRY
AND MEASURES

 Revision Guide
Page 75

Hint

You can answer lots of question about regular polygons by finding the size of an exterior angle.

The size of an exterior angle in a regular *n*-sided polygon is 360° ÷ *n*

Explore

The size of one exterior angle in a regular polygon is **inversely proportional** to the number of sides.

---

4   The size of each interior angle of a regular polygon with *n* sides is 140°

Work out the size of each interior angle of a regular polygon with 2*n* sides.

 Scan this QR code for a video of this question being solved!

.... °

(Total for Question 4 is 4 marks)

   Turn to page 130 for complete worked solutions to the questions on this page.

5   Zoe asked a group of 25 friends to complete two puzzles.

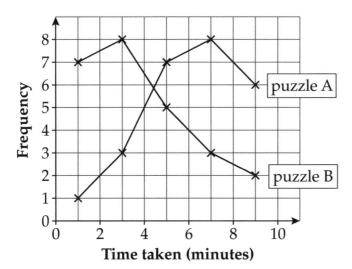

The frequency polygon shows the times taken by each of her 25 friends to complete each puzzle.

Which puzzle was harder?
Give a reason for your answer.

....

....

....

(Total for Question 5 is 2 marks)

PROBABILITY & STATISTICS

Revision Guide
Pages 121, 122

Problem solved!

Always give **evidence** when you are comparing two distributions. For example, you could compare the number of friends who solved each puzzle in less than 2 minutes.

Watch out!

Don't waste time writing long answers. Compare some values for each distribution, then write a short conclusion.

Explore

Frequency polygons are usually used to compare grouped continuous data. If you tried to draw two histograms on the same graph it would be very confusing.

Turn to page 131 for complete worked solutions to the questions on this page.

PROBABILITY
& STATISTICS

 Revision Guide
Pages 112, 119

Hint

Use the smallest value and the range to calculate the largest value.

Hint

Use the upper quartile and the interquartile range to calculate the lower quartile.

Hint

You need to use a sharp pencil and a ruler to draw a neat diagram.

6 Here is some information about the masses, in kg, of 15 cabbages.

Smallest	1.0
Median	1.4
Upper quartile	1.6
Range	0.85
Interquartile range	0.4

On the grid draw a box plot to show this information.

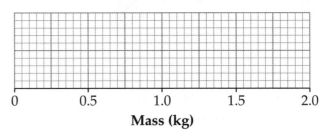

Mass (kg)

(Total for Question 6 is 2 marks)

Turn to page 131 for complete worked solutions to the questions on this page.

7   The *n*th term of sequence A is $3n - 2$
The *n*th term of sequence B is $10 - 2n$

Sally says there is only one number that is in both sequence A and sequence B.

Is Sally right?
You must explain your answer.

√xy² **ALGEBRA**

Revision Guide
Pages 22, 23

Problem solved!

Write out the first few terms of each sequence. This should show you what is going on. Remember to write a conclusion explaining your answer.

Watch out!

Don't write
$3n - 2 = 10 - 2n$
and try to solve an equation. The value of *n* might be **different** in each sequence.

Hint

Remember you can only use the general term for a sequence with **positive** values of *n*. *n* = 0 or negative values of *n* don't give you terms in the sequence.

(Total for Question 7 is 2 marks)

Turn to page 131 for complete worked solutions to the questions on this page.

RATIO AND PROPORTION

 Revision Guide Page 69

Problem solved!

Part **(a)** says "prove" so use algebra, and make sure you write a conclusion saying what you have proved.

Hint

Start by writing proportionality formulae for *x* and *y*, and for *y* and *z*:

$x = ky$

$y = \dfrac{j}{z}$

Watch out!

But make sure you use **different letters** for the constant of proportionality in each formula.

 **Explore**

If two quantities are inversely proportional their product will be constant.

---

8 *x* is directly proportional to *y*

   y is inversely proportional to *z*

   (a) Prove that *z* is inversely proportional to *x*

   (3)

   When $x = 40$, $z = 0.2$

   (b) Work out the value of *z* when $x = 16$

   $z = $ ....

   (2)

   (Total for Question 8 is 5 marks)

Turn to page 131 for complete worked solutions to the questions on this page.

9 On the grid below, show by shading the region defined by the inequalities

$$y > 1 \qquad y < 2x - 2 \qquad y < 6 - x \qquad x > 0$$

Mark this region with the letter R.

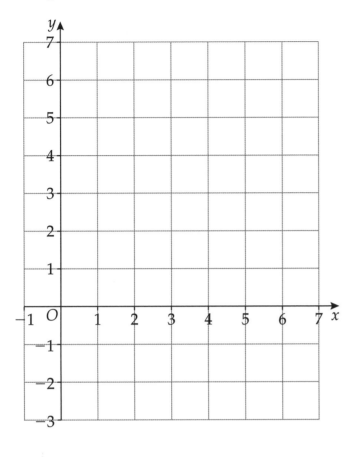

(Total for Question 9 is 4 marks)

 ALGEBRA

Revision Guide Page 41

Hint

Swap each inequality for an equals sign, then draw the straight lines for each equation. Draw arrows to show which side of the line you are interested in. For $y <$ ... you are interested in the area **below** the line.

Hint

For $\geq$ and $\leq$ you use a solid line. For $>$ and $<$ you use a dashed line.

LEARN IT!

Lines of the form $y = \square$ are horizontal. Lines of the form $x = \square$ are vertical.

Explore

Pick a point inside your region and check that it satisfies all the inequalities. Pick a point outside and check that it does not.

Turn to page 132 for complete worked solutions to the questions on this page.

GEOMETRY AND MEASURES

Revision Guide
Pages 73, 74

Problem solved!

You can draw extra lines on the diagram to help you. Try continuing *ED* until it meets the line *BC*. Label the point where it meets *F* so you can show your working clearly.

Hint

Write any angles you work out on your diagram as you go.

Watch out!

You must give reasons for **each step** of your working.

Watch out!

Unless it says so in the question, the diagram is **not accurate** so don't measure any angles on the diagram. You can use the diagram to check that your answer **makes sense**.

---

10 The diagram shows a pentagon *ABCDE*.

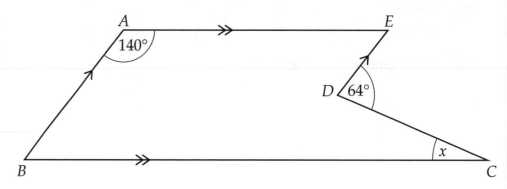

AE is parallel to *BC*.
BA is parallel to *DE*.

Angle *EDC* = 64°
Angle *BAE* = 140°

Work out the size of the angle marked *x*.
You must give reasons for your answer.

$x = $ .... °

(Total for Question 10 is 4 marks)

Turn to page 132 for complete worked solutions to the questions on this page.

11 Icetown makes fridges.

The probability that an Icetown fridge will have an electrical fault is 0.02
The probability that an Icetown fridge will have a mechanical fault is 0.05

(a) Complete the decision tree diagram.

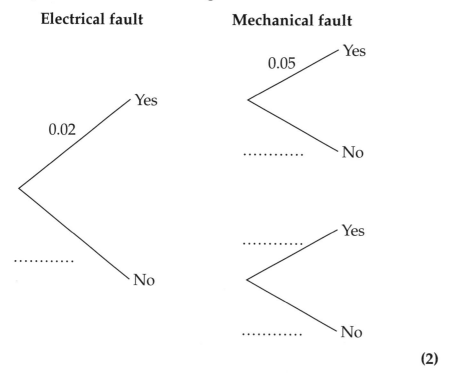

Electrical fault   **Mechanical fault**

0.02 ——— Yes

0.05 ——— Yes

············ ——— No

············ ——— No

············ ——— Yes

············ ——— No

(2)

Coolbox also makes fridges.

The probability that a Coolbox fridge will have **no** electrical fault and **no** mechanical fault is 0.93

Janet wants to buy a fridge with the least risk of any fault.

(b) Which make of fridge should Janet buy, an Icetown fridge or a Coolbox fridge?

(3)
(Total for Question 11 is 5 marks)

PROBABILITY & STATISTICS

Revision Guide
Page 127

LEARN IT!

At each branch the probabilities add up to 1

LEARN IT!

In a tree diagram you multiply along the branches to find the probability of each outcome:

MULTIPLY ALONG THE BRANCHES + ADD UP THE OUTCOMES

Watch out!

You need to **answer the question**. Calculate the probability that an Icetown fridge has no faults, then **write a conclusion**.

Explore

The probability that two events **both** occur must be less than each separate probability. You can use this to check your answer.

Turn to page 132 for complete worked solutions to the questions on this page.

33

 **ALGEBRA**

Revision Guide
Page 50

Hint

Always use brackets when substituting into a function.

Watch out!

When calculating a **composite function** the order is important. fg(−3) means work out g(−3) first, then use that answer as the **input** for f.

Hint

To find the inverse of a function:

- write it as $y = \ldots$
- rearrange to make x the subject
- swap y for x
- write in the form $f^{-1}(x) = \ldots$

Explore

Choose a value of x and apply the function f. Then apply f^{-1} using your answer to part (**d**). You have just worked out $f^{-1}f(x)$. What do you notice?

12 The functions f and g are defined as

$$f(x) = \tfrac{1}{2}x + 4$$

$$g(x) = \frac{2x}{x+1}$$

(a) Work out f(6)

....
(1)

(b) Work out fg(−3)

....
(2)

(c) g(a) = −2

Work out the value of a.

$a = $ ....
(2)

(d) Express the inverse function f^{-1} in the form $f^{-1}(x) = \ldots$

$f^{-1}(x) = $ ....
(3)

(Total for Question 12 is 8 marks)

Turn to page 132 for complete worked solutions to the questions on this page.

13

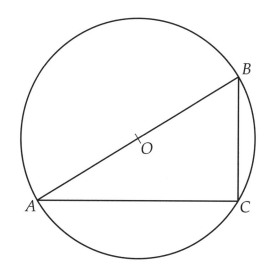

The diagram shows a circle centre O with diameter AB.

C lies on the circumference of the circle.

Prove that the angle in a semicircle is a right angle.

GEOMETRY AND MEASURES

 Revision Guide Page 105

LEARN IT!

The angle at the centre of a circle is twice the angle at the circumference:

You can use this fact in your proof.

Watch out!

A proof isn't complete until you have written a conclusion. Finish by stating what you have proved.

Explore

Draw some different diameters on the circle. Use the corner of a piece of paper to demonstrate this circle fact:

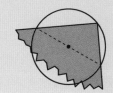

(Total for Question 13 is 4 marks)

Turn to page 133 for complete worked solutions to the questions on this page.

% RATIO AND PROPORTION

 Revision Guide
Page 64

Hint

To find the multiplier for a percentage decrease, subtract the decrease from 100 then divide by 100:

$$\frac{100 - 20}{100} = 0.8$$

LEARN IT!

Use this formula to find the amount after n years:

$$\binom{\text{starting}}{\text{amount}} \times (\text{multiplier})^n$$

Hint

Find the amount after 1 year, 2 years, 3 years and so on. Remember to write the value of n as your final answer.

---

14 The value of a van depreciates at the rate of 20% per year.
Gary buys a new van for £27 500
After n years the value of the van is £11 264

Find the value of n.

$n = $ ....

(Total for Question 14 is 2 marks)

Turn to page 133 for complete worked solutions to the questions on this page.

15 A car accelerates from 0 metres per second to 60 metres per second in 20 seconds.
It then travels at a constant speed of 60 metres per second for 30 seconds.

The speed–time graph shows this information.

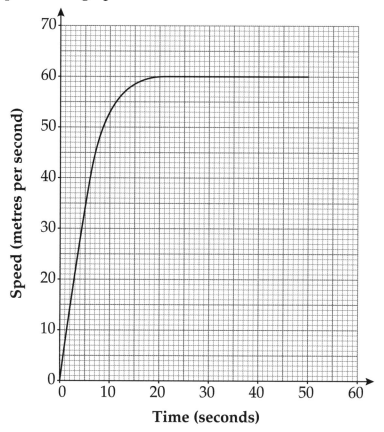

Work out an estimate for the distance the car travelled in these 50 seconds.

 **ALGEBRA**

Revision Guide
Pages 55, 56

LEARN IT!

The area under a speed–time (or velocity–time) graph is the distance travelled.

Hint

Divide the area under the graph into triangles, rectangles and trapeziums to work out your estimate. Draw these shapes on the graph to show your working.

Watch out!

Use the **scale** when calculating areas – don't count squares.

LEARN IT!

Area = $\frac{1}{2}(a + b)h$

Explore

Each of your shapes will be completely **under** the graph, so your estimate will be **smaller** than the actual distance travelled.

.....................................km

(Total for Question 15 is 3 marks)

Turn to page 133 for complete worked solutions to the questions on this page.

RATIO AND
PROPORTION

 Revision Guide
Page 64

Watch out!

You don't need to round your answers.

LEARN IT!

Learn your percentage increase and decrease multipliers:

20% decrease: × 0.8

30% decrease: × 0.7

Problem solved!

For part **(b)** you need to keep track of the amounts at each stage. Don't try and take any short cuts. Work out the amount after 6 hours, 12 hours and 18 hours.

Watch out!

Remember to use the output from the previous calculation at each stage.

---

16 The mass of substance A decreases at the rate of 20% every 6 hours.

500 grams of substance A is put into a dish.

(a) Work out the mass of substance A in the dish at the end of 12 hours.

.... grams

(3)

500 grams of a different substance B is placed in a flask.
The mass of this substance decreases at a rate of 30% every 6 hours.

At the end of 6 hours, 500 grams more of substance B is added to the flask.
At the end of a further 6 hours, another 500 grams of substance B is added to the flask.

(b) Work out the mass of substance B in the flask at the end of 18 hours.

.... grams

(3)

(Total for Question 16 is 6 marks)

Turn to page 133 for complete worked solutions to the questions on this page.

17 Clive wants to estimate the number of fish in a pond.

Clive catches 50 fish from the pond.
He marks each fish with a dye.
He then puts the fish back in the pond.

The next day, Clive catches 40 fish from the pond.
8 of these fish have been marked with the dye.

Work out an estimate for the number of fish in the pond.

LEARN IT!

$N = \dfrac{Mn}{m}$

N = total population size
M = number of fish marked then released
n = size of recapture sample
m = number of marked fish in recapture sample.

Explore

When scientists use the capture-recapture method, they have to leave enough time between each sample. The method only works if the marked animals have had time to be randomly distributed amongst the population.

Scan this QR code for a video of this question being solved!

.... fish
(Total for Question 17 is 2 marks)

Turn to page 134 for complete worked solutions to the questions on this page.

√xy² **ALGEBRA**

Revision Guide
Pages 26, 27

Hint

You need to work out the coordinates of points B, C and A. Write them on the diagram once you have found them.

Hint

The question tells you that AB = BC, so B is the **midpoint** of AC.

Hint

You can find the point where a graph crosses the x-axis by substituting y = 0

LEARN IT!

To find the equation of a line through two points, A and D:

• find the gradient of AD
• choose one point and substitute the coordinates into y = mx + c
• solve to find c
• write out the equation.

Watch out!

The values of m and c in the equation of a line don't have to be whole numbers.

18

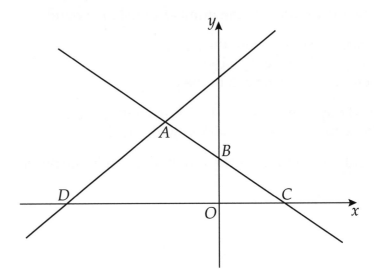

In the diagram, ABC is the line with equation $y = -\frac{1}{2}x + 5$

$AB = BC$

D is the point with coordinates $(-13, 0)$.

Find an equation of the line through A and D.

....
(Total for Question 18 is 5 marks)

Turn to page 134 for complete worked solutions to the questions on this page.

19 (a) The table and histogram show some information about the mass, in grams, of some batteries.

Use the table to complete the histogram.

Mass (m grams)	Frequency
$30 < m \leqslant 40$	4
$40 < m \leqslant 50$	6
$50 < m \leqslant 65$	15
$65 < m \leqslant 80$	9
$80 < m \leqslant 100$	4

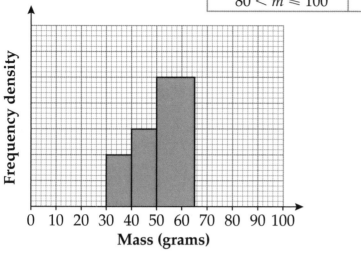

(2)

The histogram shows information about the lifetime of some batteries.

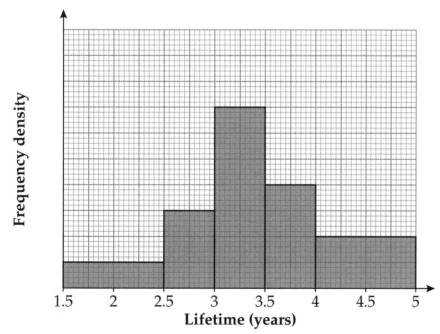

(b) Two of the batteries had a lifetime of between 1.5 and 2.5 years. Find the total number of batteries.

(3)

(Total for Question 19 is 5 marks)

 PROBABILITY & STATISTICS

 Revision Guide Page 120

Hint

On a histogram, the **area** of a bar is proportional to the frequency.

Problem solved!

In part **(b)** there is no scale on the vertical axis, so you have to use area. Work out the total area of all the bars by counting small graph squares. Then use the fact that the first bar represents 2 batteries to work out how many small squares represent 1 battery.

Explore

You can use **interpolation** to estimate values within bars. If you need to include X% of the area of a bar, you can choose a value X% of the way along the width of that bar.

Turn to page 134 for complete worked solutions to the questions on this page.

RATIO AND PROPORTION

Revision Guide
Page 64

Problem solved!

Use $900 = 1200 \times k^2$ to find the rate of decrease in **one year, k**

Watch out!

Check that your answer makes sense. The number of bees is decreasing, so your answer must be **more than 1200**

Hint

Round your answer to a sensible degree of accuracy.

Hint

Write words with your working to show what you are calculating at each stage.

---

20 Rhys has a beehive.
The number of bees in the beehive is decreasing.

Rhys counts the number of bees in the hive at the start of week 5
He counts the number of bees in the hive at the start of week 7
Here are his results.

week	number bees
5	1200
7	900

Assuming that the population of bees is decreasing exponentially, how many bees were there at the start of week 2?
You must show your working.

.... bees
(Total for Question 20 is 4 marks)

TOTAL FOR PAPER IS 80 MARKS

Turn to page 134 for complete worked solutions to the questions on this page.

Paper 3: Calculator
Time allowed: 1 hour 30 minutes

Revision Guide
Pages 112, 122

1   The heights (in cm) of 13 girls and 13 boys were recorded.

The back-to-back stem-and-leaf diagram gives this information.

girls						boys			
			9	8	14				
			4	2	15	7	9		
8	4	4	2	0	16	2	6	8	9
	9	5	3	0	17	0	3	4	6	6
					18	1	4		

KEY:

8 | 14 represents a height of 148 cm for girls

15 | 7 represents a height of 157 cm for boys

Compare the distribution of the heights of the girls and the distribution of the heights of the boys.

(Total for Question 1 is 3 marks)

Turn to page 135 for complete worked solutions to the questions on this page.

Problem solved!

Always give **evidence** when you are comparing two distributions. Try to compare one average and one measure of spread. Make sure your statements refer to the **context** of the question. You need to talk about **heights**.

Hint

For back-to-back stem-and-leaf diagrams, it is a good idea to compare the **range** or the **interquartile range** and the **median**.

Watch out!

When reading data values from a stem-and-leaf diagram, make sure you use the full data value, not just the 'leaf'.

Watch out!

Don't waste time writing long-winded answers. Compare values for each distribution, then write a short conclusion.

RATIO AND PROPORTION

 Revision Guide
Page 63

Watch out!

This is a **reverse percentages** question. You are given the amount **after** the increase and you need to find the **original amount**.

Hint

Here are two possible strategies:

1. Find the multiplier for a 25% increase, then **divide** 5950 by the multiplier.

2. Divide 5950 by 125 to find 1%, then multiply by 100 to find 100%.

Hint

Increase your answer by 25% and **check** that you get 5950

2   A number is increased by 25% to get 5950

What is the number?

....

(Total for Question 2 is 2 marks)

Turn to page 135 for complete worked solutions to the questions on this page.

3 Caroline is making some table decorations.
Each decoration is made from a candle
and a holder.

Caroline buys some candles and some
holders each in packs.

There are 30 candles in a pack of candles.
There are 18 holders in a pack of holders.

candle and
holder

Caroline buys exactly the same number of
candles and holders.

(a) How many packs of candles and how many packs of holders
does Caroline buy?

.... packs of candles

.... packs of holders
(3)

Caroline uses all her candles and all her holders.

(b) How many table decorations does Caroline make?

.... table decorations
(1)

(Total for Question 3 is 4 marks)

NUMBER

  Revision Guide
Page 1

Hint

The number of candles
must be a multiple
of 30. The number
of holders must be a
multiple of 18. So you
need to find the LCM
of 30 and 18

Watch out!

Caroline buys the same
number of candles
and holders. But she
doesn't buy the same
number of packs.

Problem solved!

Keep track of your
working. You could
draw a table showing
the number of candles
and holders for
different numbers of
packs. Circle the first
numbers that match,
and remember to write
the number of **packs**
as your answer.

Turn to page 135 for complete worked solutions to the questions on this page.

ALGEBRA

 GEOMETRY AND MEASURES

Revision Guide
Pages 37, 80

Hint

Work out expressions for AD and AB in terms of x. Write these expressions on the large rectangle.

Problem solved!

Write expressions for the perimeter of each rectangle and use these expressions to write an inequality involving x.

Watch out!

You need to add together **four** lengths to get the perimeter:
Perimeter = 2 × length + 2 × width

 **Explore**

You can't have a negative length. You should write your final inequality in the form
$0 \le x < \square$

4 Here are two metal plates in the shape of rectangles.

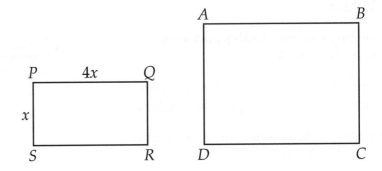

In the diagram, all the measurements are in cm.

The length of AD is twice the length of PS.
The length of AB is 5 cm more than the length of PS.

Find the range of values of x for which the perimeter of the rectangle ABCD is greater than the perimeter of the rectangle PQRS.

....

(Total for Question 4 is 4 marks)

Turn to page 135 for complete worked solutions to the questions on this page.

5 Bill wants to compare the heights of pine trees growing in sandy soil with the heights of pine trees growing in clay soil.

The scatter diagram gives some information about the heights and the ages of some pine trees.

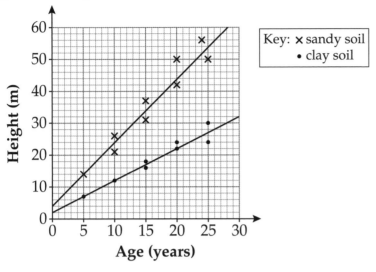

A pine tree growing in clay soil is 18 years old.

(a) Find an estimate for the height of this tree.

.....................................m

(1)

A pine tree is growing in sandy soil.

(b) Work out an estimate for how much the height of this tree increases in a year.

.....................................m

(2)

(c) Compare the rate of increase of the height of trees growing in clay soil with the rate of increase of the height of trees growing in sandy soil.

....

....

....

(2)

(Total for Question 5 is 5 marks)

PROBABILITY & STATISTICS

Revision Guide Page 114

Hint

Always draw lines on your graph to show any values you are reading off.

Hint

The gradient of the line of best fit tells you how much the height increases for each additional year of growth.

LEARN IT!

Draw a triangle on the graph to find the gradient:

$$\text{Gradient} = \frac{\text{Distance up}}{\text{Distance across}}$$

Hint

The gradient of the line of best fit tells you the **rate of change** of the height. The steeper the gradient, the faster the trees are growing.

Explore

How confident are you about your conclusion for part **(c)**? How could a scientist increase the accuracy of this statement?

Turn to page 136 for complete worked solutions to the questions on this page.

GEOMETRY AND MEASURES

Revision Guide
Page 83

LEARN IT!

Volume of a cylinder = $\pi r^2 h$

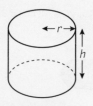

Watch out!

You are given the **diameter** of each cylinder. The formula for the volume of a cylinder uses the **radius**, so divide by 2

Problem solved!

Plan your strategy. You will need to find the volume of the can. Once the soup has been poured into the saucepan it forms a cylinder of diameter 12 cm with the **same volume**.

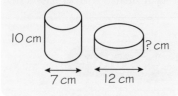

10 cm   ? cm

7 cm   12 cm

---

6 A can of soup is a cylinder with a diameter 7 cm.
The can is 10 cm high.
The can is full of soup.

The soup is poured into a saucepan.
The saucepan is a cylinder with a diameter 12 cm.

Work out the depth of the soup in the saucepan.
Give your answer correct to 1 decimal place.

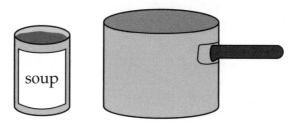

soup

Scan this QR code for a video of this question being solved!

....................................cm
(Total for Question 6 is 3 marks)

Turn to page 136 for complete worked solutions to the questions on this page.

7   The diagram shows a chocolate bar in the shape of a triangular prism.

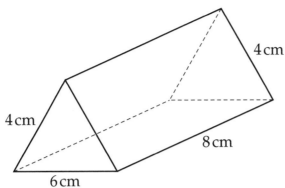

On the centimetre grid, draw a plan of the chocolate bar.

(Total for Question 7 is 2 marks)

GEOMETRY AND MEASURES

Revision Guide Page 87

LEARN IT!

The plan is the view from above the shape.

Hint

It doesn't matter which way round your shape is. But you must draw it accurately, so use a ruler and a sharp pencil.

Hint

Draw a line on your plan to show the change in height at the top edge of the prism.

Explore

The cross-section is an isosceles triangle, so you know that the top edge will be in the **middle** of your plan.

Turn to page 136 for complete worked solutions to the questions on this page.

PROBABILITY & STATISTICS

Revision Guide
Pages 123, 124

Hint

There are 16 equally likely outcomes. Work out the number of outcomes with a total of 7 or more and divide by 16 to find the probability that the player wins.

Hint

The **expected** number of wins is P(win) × 100

Hint

Show all your working to justify your answer, and make sure you write a conclusion explaining whether Megan can expect to make a profit.

Explore

Your working won't show whether Megan will **definitely** make a profit. If you flip a fair coin 100 times you can **expect** to get 50 heads, but it is unlikely that you will get **exactly** 50 heads.

---

8   Megan is planning a game to raise money for charity.
She is going to use a fair spinner.

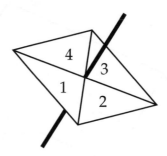

Megan spins the spinner twice.
The score is the sum of the numbers the spinner lands on.

(a) Complete the table to show the possible scores.

2nd spin 1st spin	1	2	3	4
1				
2	3		5	
3		5		
4				8

(1)

Here are the rules for Megan's game:

Pay 50p to spin the spinner twice.
When the score is 7 or more, get £1.50.

(b) If this game is played 100 times should Megan expect to make a profit?

(5)
(Total for Question 7 is 6 marks)

Turn to page 136 for complete worked solutions to the questions on this page.

9 $x^2 - x - 6 \leqslant 0$

Show the solution to this inequality on the number line below.

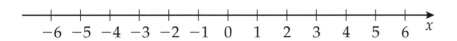

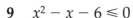 **ALGEBRA**

Revision Guide
Page 38

Hint

This is a **quadratic inequality**. You need to **factorise** the left-hand side.

Hint

Sketch the graph of $y = x^2 - x - 6$. The inequality is $\leqslant 0$ so you are interested in the values of x where the graph is **below or on** the x-axis.

Watch out!

When you are marking inequalities on a number line, use a solid dot (●) for $\leqslant$ and $\geqslant$ and a circle (○) for $<$ and $>$.

Explore

You can use set notation to describe solutions to inequalities. $\{x: x < 20\}$ means 'the set of values of x such that x is less than 20'.

(Total for Question 9 is 4 marks)

Turn to page 137 for complete worked solutions to the questions on this page.

**GEOMETRY
AND MEASURES**

 Revision Guide
Page 84

 LEARN IT!

A sector with angle *x*
and radius *r* has arc
length:

$$\frac{x}{360°} \times 2\pi r$$

Watch out!

The edging needs to
go around the whole
perimeter. Calculate the
arc length, then add on
two lots of the radius.

Hint

Toby has to buy a
whole number of
lengths. Divide the
perimeter by 1.75
then **round up** to the
nearest whole number.

---

10 The diagram shows a pond.

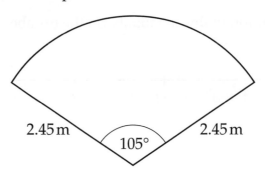

The pond is in the shape of a sector of a circle.

Toby is going to put edging on the perimeter of the pond.

Edging is sold in lengths of 1.75 metres.
Each length of edging costs £3.49.

Work out the total cost of edging Toby needs to buy.

£ ....

(Total for Question 10 is 5 marks)

Turn to page 137 for complete worked solutions to the questions on this page.

11 Jade makes a blackcurrant drink by mixing orange concentrate with water.

She mixes $15\,\text{cm}^3$ of blackcurrant concentrate with $250\,\text{cm}^3$ of water.

The density of blackcurrant concentrate is $1.20\,\text{g/cm}^3$.
The density of water is $1.00\,\text{g/cm}^3$.

Work out the density of Jade's blackcurrant drink.
Give your answer correct to 2 decimal places.

RATIO AND PROPORTION

Revision Guide
Page 66

LEARN IT!

Draw the formula triangle for density at the top of your working:

Hint

You need to find the **total mass** and the **total volume** of Jade's drink to calculate the density.

Explore

The units of density can give you a clue about how to calculate it. The units are grams/cm³ and to find density you work out $\dfrac{\text{mass}}{\text{volume}}$

Scan this QR code for a video of this question being solved!

.....................................g/cm^3
(Total for Question 11 is 3 marks)

GEOMETRY
AND MEASURES

Revision Guide
Page 105

LEARN IT!

If you see a **tangent**
and a **chord** in a circle
question, you might
be able to use the
**alternate segment
theorem.**

Hint

Your answer to part
(a) can be used in your
solution to part **(b)**.

Hint

Write any angles you
work out on your
diagram as you go.
Remember, you still
need to write down
reasons for each step
of your working.

12

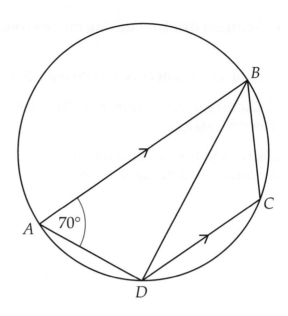

A, *B*, *C* and *D* are points on a circle.
AB is a diameter of the circle.
DC is parallel to *AB*.
Angle *BAD* = 70°

(a) Show that angle *BDC* = 20°

You must give reasons for your working.

(3)

The tangent to the circle at *D* meets the line *BC* extended at *T*.

(b) Calculate the size of angle *BTD*.

o
....

(3)
(Total for Question 12 is 6 marks)

Turn to page 137 for complete worked solutions to the questions on this page.

13 The area covered by the Pacific Ocean is $1.6 \times 10^8 \text{ km}^2$.
The area covered by the Arctic Ocean is $1.4 \times 10^7 \text{ km}^2$.

(a) Write 1.6×10^8 as an ordinary number.

....
(1)

The area covered by the Pacific Ocean is k times the area covered
by the Arctic Ocean.

(b) Find, correct to the nearest integer, the value of k.

$k = $ ....
(2)
(Total for Question 13 is 3 marks)

NUMBER

Revision Guide
Page 8

LEARN IT!

For a number in
standard form:

• the first part is a
  number ⩾1 and <10
• the second part is a
  power of 10

Hint

Count decimal places
to convert between
ordinary numbers and
standard form.

LEARN IT!

Numbers larger than 10
have a **positive** power
of 10

Numbers less than 1
have a **negative** power
of 10

Explore

Use the ×10ˣ key to
enter a number in
standard form on your
calculator. Investigate
how the ENG function
affects answers in
standard form.

Turn to page 138 for complete worked solutions to the questions on this page.

√xy² **ALGEBRA**

 Revision Guide
Pages 23, 24

Hint

You can sometimes spot the rule for a sequence of patterns by looking at how the patterns are formed. Each rectangle is twice as wide as it is tall.

Hint

The next pattern in the sequence will be 4 squares tall and 8 squares wide.

Problem solved!

The easiest way to explain your answer to part **(c)** is to show some working. Set your expression for the nth term equal to 200, and solve the equation to find n. If n is an integer, then 200 is a term in the sequence.

14 Here is a sequence of patterns made from centimetre squares.

Pattern number 1   Pattern number 2

Pattern number 3

(a) Write down the number of centimetre squares used in pattern number 4.

....
(1)

(b) Find an expression, in terms of n, for the number of centimetre squares used in pattern number n.

....
(2)

(c) Alex says there is a pattern in this sequence which is made from 200 centimetre squares.

Is Alex correct?
Show your working.

....
(2)
(Total for Question 14 is 5 marks)

Turn to page 138 for complete worked solutions to the questions on this page.

15 The histogram shows information about the lifetime of some electrical components.

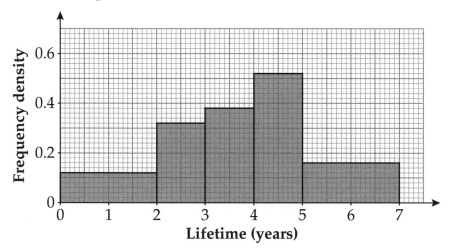

Work out the proportion of the components with a lifetime of between 1 and 6 years.

....

(Total for Question 15 is 4 marks)

 PROBABILITY & STATISTICS

Revision Guide Page 120

LEARN IT!

On a histogram, **area** is proportional to frequency.

Hint

Compare the area of the bars between 1 and 6 years with the total shaded area. Use the scale to calculate areas rather than just counting squares.

Hint

You can give your proportion as a fraction or decimal between 0 and 1, or as a percentage between 0% and 100%.

Explore

Your answer will be an **estimate**. You are assuming that the lifetimes of the components in the 0–2 class and the 5–7 class are **evenly distributed** within those classes.

Turn to page 138 for complete worked solutions to the questions on this page.

√xy² **ALGEBRA**

Revision Guide
Page 54

Hint

The rate of cooling at a given time is the **gradient** of the curve at that point. To find the gradient of a curve you have to draw a tangent to the curve.

LEARN IT!

The **tangent** to a curve at a point is the straight line that just touches the curve at that point.

Watch out!

Your tangent will be more accurate if you use a sharp pencil. Mark the point on the curve, then slide your ruler until it **just** touches that point.

Explore

When you work out the gradient you are dividing a temperature (°C) by a time (mins), so the units of the gradient are °C/min.

16 Hot drinks are served at a temperature of 70°C.

The graph shows the temperature of a hot drink as it cools in a china mug from the time it is served.

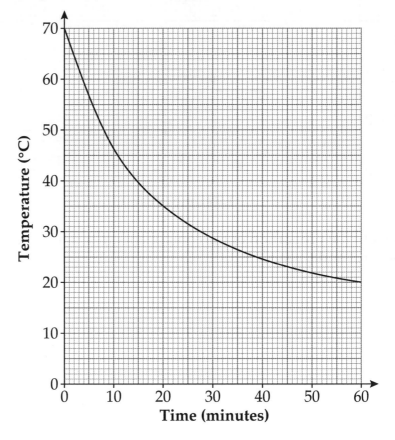

Work out the rate of cooling of the drink at time 20 minutes.

Scan this QR code for a video of this question being solved!

.....................................°C per minute

(Total for Question 16 is 3 marks)

Turn to page 138 for complete worked solutions to the questions on this page.

17 Expand and simplify $(x - 1)(x + 2)(x + 3)$

Scan this QR code for a video of this question being solved!

....
(Total for Question 17 is 3 marks)

  ALGEBRA

 Revision Guide Page 17

Watch out!

You might have to expand the product of three linear brackets in your exam.

Hint

Expand the last two brackets to get (linear) × (quadratic). Then multiply **each term** in the linear expression by the **whole** quadratic expression.

Explore

All three brackets contain an x term, so the highest power of x in your final answer will be x^3.

Turn to page 139 for complete worked solutions to the questions on this page.

 NUMBER

% RATIO AND PROPORTION

 Revision Guide
Pages 10, 65

Hint

This is a question on **upper and lower bounds**. You need to choose the correct bounds for each measurement to work out the greatest possible speed of the car.

Hint

Write down the UB and LB of each rounded measurement before doing any calculations.

Watch out!

You need to convert seconds to hours and metres to km, so you can work out the greatest possible speed in km/h.

Watch out!

The accuracy affects the upper and lower bounds. The time is measured to the nearest second so the actual value could be 0.5 s higher or lower. The distance is measured to the nearest 10 m so the actual value could be 5 m higher or lower.

18 A car is driven through a tunnel in 89 seconds, correct to the nearest second.

The tunnel is 2460 m long, correct to the nearest 10 metres.

The average speed limit in the tunnel is 100 km/h.

Could the average speed of this car have been greater than 100 km/h? You must show your working.

Scan this QR code for a video of this question being solved!

(Total for Question 18 is 4 marks)

Turn to page 139 for complete worked solutions to the questions on this page.

19 Here is a speed-time graph.

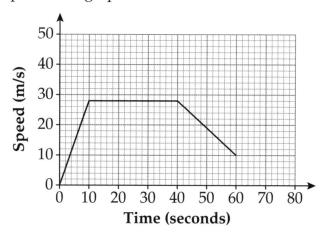

The diagram gives information about the speed of a car along a
road for 60 seconds.
The speed limit for the road is 110 kilometres per hour.

(a) Does the speed of the car exceed the speed limit?

(3)

(b) Work out the acceleration of the car during the first 10 seconds.
    You must write units with your answer.

....

(3)

(Total for Question 19 is 6 marks)

RATIO AND PROPORTION

Revision Guide
Pages 55, 65

Hint

For part **(a)** you need
to read the maximum
speed in m/s off the
graph, then convert to
km/h.

Hint

You can convert from
m/s to km/h in two
stages:

$$m/s \xrightarrow{\times 3600} m/h \xrightarrow{\div 1000} km/h$$

Hint

The **gradient** of a
speed-time graph gives
you the acceleration.

LEARN IT!

The units of acceleration
are m/s/s or m/s².

Turn to page 139 for complete worked solutions to the questions on this page.

61

 **ALGEBRA**

 Revision Guide
Page 45

Hint

If you have to use an iteration formula in your exam you will be given the formula, and a starting value.

Watch out!

When working out values for an iteration, write down all the digits off your calculator display.

Explore

You can do iteration quickly on your calculator. Press

 , then
enter the calculation, replacing x_n with the ANS function.

Press $=$ for each iteration.

20 (a) Show that the equation $x^3 - 4x + 1 = 0$ can be rearranged to give

$$x = \frac{x^3}{4} + \frac{1}{4}$$

(1)

(b) Starting with $x_0 = 1$, use the iteration

$$x_{n+1} = \frac{x^3{}_n}{4} + \frac{1}{4}$$

to calculate the values of x_1, x_2 and x_3.

(3)

(c) Explain what the values of x_1, x_2 and x_3 represent.

....

....

(1)

(Total for Question 20 is 5 marks)

TOTAL FOR PAPER IS 80 MARKS

Turn to page 139 for complete worked solutions to the questions on this page.

Paper 1: Non-calculator
Time allowed: 1 hour 30 minutes

1 Expand and simplify $(x + 4)(x + 6)$

....

(Total for Question 1 is 2 marks)

 ALGEBRA

 Revision Guide
Page 17

Hint

You can use a grid to expand double brackets:

	x	$+4$
x		$+4x$
$+6$		

Complete all four terms then add them together.

Explore

Check your answer by substituting a value for x into your expression. When $x = 1$ the original expression is $(1 + 4)(1 + 6) = 5 \times 7 = 35$. Does your final expression give the same value?

Turn to page 140 for complete worked solutions to the questions on this page.

RATIO AND PROPORTION

 Revision Guide Page 61

Hint

Write words with your working to explain what you are working out at each stage.

Hint

Use mental strategies to save time. The total distance Mark drives in one week is 5 × 24 = 5 × 20 + 5 × 4 = 100 + 20 = 120 miles.

Problem solved!

You need to work out the total cost of the diesel Mark would use in one week. Then compare this with the cost of the weekly train pass and write a conclusion.

Hint

If you need to convert between metric units (like litres) and imperial units (like gallons) in your exam, you will be given the conversion with the question.

2 Mark works for 5 days each week.

Mark can travel to work by car or train.

By car

He travels a total distance of 24 miles each day

His car travels 30 miles per gallon

Diesel costs £1.50 per litre

By train

Weekly pass costs £25.75

---

1 gallon = 4.5 litres

Is it more expensive if he uses his car or the train?

You must show your working.

(Total for Question 2 is 4 marks)

Turn to page 140 for complete worked solutions to the questions on this page.

3 (a) Complete the table of values for $y = x^3 - 4x$

x	-3	-2	-1	0	1	2	3
y			3	0			15

(2)

 **ALGEBRA**

 Revision Guide
Page 29

Hint

To complete the table, substitute each value of x into the right-hand side of the equation and work out the corresponding value of y.

(b) On the grid, draw the graph of $y = x^3 - 4x$ from $x = -3$ to $x = 3$

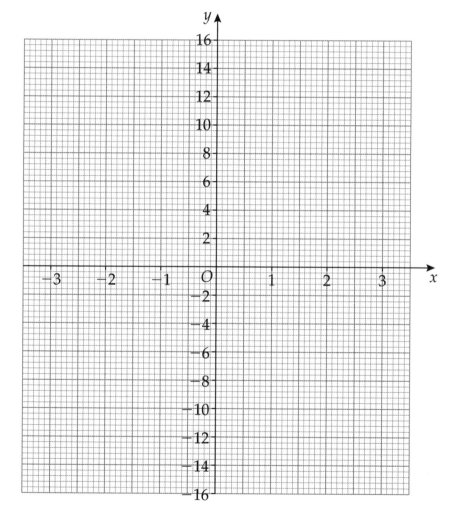

(2)

(c) Use your graph to find estimates of the solutions to the equation
$x^3 - 4x = 2$

....

(2)

(Total for Question 3 is 6 marks)

Watch out!

When you cube a negative number the answer is negative:
$(-3)^3 = -3 \times -3 \times -3$
$\qquad = -27$

Hint

This is an example of a **cubic graph**. The graph should be a smooth curve. If one of your points doesn't follow the pattern check your working.

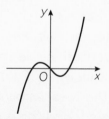

Explore

These solutions are based on reading off a graph. The accuracy of the solutions will depend on the accuracy of your graph.

Turn to page 140 for complete worked solutions to the questions on this page.

GEOMETRY AND MEASURES

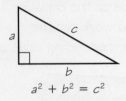 Revision Guide
Pages 76, 83

LEARN IT!

Use Pythagoras' theorem to find the length of the diagonal:

$a^2 + b^2 = c^2$

LEARN IT!

You need to know the square numbers up to 15^2, and their corresponding square roots.

LEARN IT!

Circumference = $\pi \times$ diameter

Hint

You can add together multiples of π. Each circle has circumference 2.5π so the total amount of wire needed for both circles is

$2.5\pi + 2.5\pi = 5\pi$

---

4   Jean makes a metal structure out of steel rope.

The diagram below shows the metal structure.

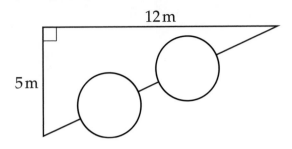

The two circles are identical.
The diameter of each circle is 2.5 metres.

Show that the total length of the steel rope used to make the metal structure is $(a + b\pi)$ metres where a and b are integers.

(Total for Question 4 is 5 marks)

Turn to page 140 for complete worked solutions to the questions on this page.

5   Simplify $3x^22y^3 \times 4xy^4 \times 2x^3y^5$

....
(Total for Question 5 is 2 marks)

 Revision Guide
Page 16

Hint

Multiply the number parts, then add the powers of each letter.

LEARN IT!

$a^m \times a^n = a^{m+n}$

Turn to page 141 for complete worked solutions to the questions on this page.

RATIO AND PROPORTION

Revision Guide
Page 62

To find the multiplier for a 20% reduction:

100% − 20% = 80%

80% ÷ 100% = 0.8

Problem solved!

You can use multipliers to solve this problem. Work out the multipliers for each percentage change, and find their product. Then work out what percentage decrease this corresponds to.

 **Explore**

An alternative strategy for this question would be to choose an amount. Imagine Asha's phone originally cost £100. Work out its price on Black Friday then write this as a percentage reduction.

6   Asha wants to buy a mobile phone.

She finds an online shop that has a sale that offers 20% off all mobile phones.

On Black Friday, the online shop reduces all sale prices by a further 30% off all mobile phones.

Asha buys a mobile phone on Black Friday.

Work out the final percentage reduction that Asha receives on the price of the mobile phone.

.... %

(Total for Question 6 is 4 marks)

Turn to page 141 for complete worked solutions to the questions on this page.

7   Here is a trapezium.

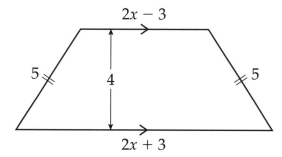

2x − 3

5

4

5

2x + 3

All the measurements are in cm.
The area of the trapezium is 18 cm².

Calculate the numerical value of the perimeter of the trapezium.

√xy² **ALGEBRA**

 GEOMETRY AND MEASURES

Revision Guide
Pages 19, 80

LEARN IT!

The formula for the
area of a trapezium will
not be given on your
exam paper:

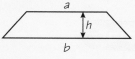

a

h

b

Area = $\frac{1}{2}(a + b)h$

Hint

You can substitute
expressions into a
formula in the same way
as numbers.

Problem solved!

Substitute $a = 2x − 3$,
$b = 2x + 3$ and $h = 4$
into the formula above
to get an expression
for the area in terms of
x. Set this equal to 18
then solve the equation
to find x.

Watch out!

The value of x does
not have to be a whole
number.

.................................cm
(Total for Question 7 is 5 marks)

Turn to page 141 for complete worked solutions to the questions on this page.

69

PROBABILITY & STATISTICS

Revision Guide
Page 125

Hint

Your blank Venn diagram should look like this:

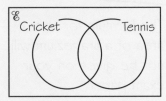

Hint

If you are drawing a Venn diagram, always start at the centre and work out. You know that 15 members play both sports, so you can write '15' in the centre of your Venn diagram.

Hint

Always check a Venn diagram by adding up all the numbers. The total should be 80

Watch out!

You don't need to simplify fractions when giving answers to probability questions.

Explore

How would the Venn diagram look different if:
• no members played both sports
• every member who played tennis also played cricket?

8   Here is some information about a cricket and tennis club.

80 people belong to the club.
35 play cricket.
50 play tennis.
15 play both cricket and tennis.

(a) Draw a Venn diagram to show this information.

(4)

One of the people who belongs to the club is chosen at random.

(b) Work out the probability that this person does not play cricket or tennis.

Scan this QR code for a video of this question being solved!

....

(2)
(Total for Question 8 is 6 marks)

Turn to page 141 for complete worked solutions to the questions on this page.

9 The distance from Caxby to Drone is 45 miles.
The distance from Drone to Elton is 20 miles.

45 miles	20 miles
Caxby	Drone	Elton

Colin drives from Caxby to Drone.
Then he drives from Drone to Elton.

Colin drives from Caxby to Drone at an average speed of 30 mph.
He drives from Drone to Elton at an average speed of 40 mph.

Work out Colin's average speed for the whole journey from Caxby to Elton.

% RATIO AND PROPORTION

Revision Guide
Page 65

Hint

Draw the formula triangle for speed at the top of your working:

$$\frac{D}{S \mid T}$$

Watch out!

The overall average speed is **not** 35 mph.

Problem solved!

You need to work out the total distance and the total time taken for the whole journey. Write words with your working to show what you are calculating at each stage.

.....................................mph
(Total for Question 9 is 3 marks)

Turn to page 142 for complete worked solutions to the questions on this page.

GEOMETRY AND MEASURES

 Revision Guide
Page 104

Hint

You don't need to use any complicated circle theorems for this question. Use one key fact about the angle between a **tangent** and a **radius**, then use angle facts about triangles and straight lines.

Hint

Write unknown angles on the diagram as you work them out.

Hint

OA and OP are both **radii** of the circle, so they are the **same length**. So triangle OAP is **isosceles**.

Watch out!

Diagrams in your exam are **not accurate** unless it says so in the question. So you can't measure any angles.

10

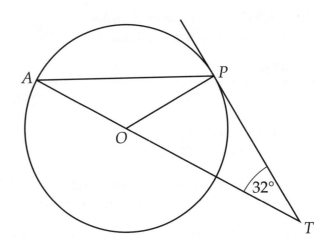

A and *P* are points on the circumference of a circle, centre *O*.
TP is a tangent to the circle.
AOT is a a straight line.
Angle *PTA* = 32°

(a) (i) State the size of angle *OPT*.

.... °

(1)

(ii) Give a reason for your answer.

....

....

....

(1)

(b) Work out the size of angle *OAP*.

.... °

(3)

(Total for Question 10 is 5 marks)

Turn to page 142 for complete worked solutions to the questions on this page.

11 (a) Write down the value of $27^{\frac{1}{3}}$

....
(1)

(b) Write down the value of $25^{-\frac{3}{2}}$

....
(2)

(c) Simplify $\dfrac{(9^x)^5}{27^x}$

....
(2)
(Total for Question 11 is 5 marks)

NUMBER

Revision Guide
Pages 2, 3

Hint

$27^{\frac{1}{3}} = \sqrt[3]{27}$

Watch out!

This is a **non-calculator
question**. You need to
know the cubes of 2, 3,
4, 5 and 10 and their
corresponding cube
roots.

Hint

If a question asks for
the 'value' of a power
you need to give your
answer as a simplified
fraction or a decimal.

Hint

$a^{-n} = \dfrac{1}{a^n}$

Problem solved!

For part **(c)** write 9
and 27 as powers of
3, then simplify. Your
final answer should be
a power of 3

Turn to page 142 for complete worked solutions to the questions on this page.

Revision Guide
Page 12

Hint

This is a question on **surds**. This is the most useful rule for working with surds:

$\sqrt{ab} = \sqrt{a} \times \sqrt{b}$

Hint

You need to find a factor of 27 that is a **square number**.

Hint

For part **(b)**, you need to write an equivalent fraction with an **integer** as the denominator. You can do this by multiplying the top and bottom of the fraction by $\sqrt{3}$. Simplify your fraction if possible.

Explore

Writing numbers as surds means you can give **exact answers** to problems.

12 (a) Express $5\sqrt{27}$ in the form $n\sqrt{3}$, where n is a positive integer.

....

(2)

(b) Rationalise the denominator of $\dfrac{21}{\sqrt{3}}$

Scan this QR code for a video of this question being solved!

....

(2)

(Total for Question 12 is 4 marks)

Turn to page 142 for complete worked solutions to the questions on this page.

13 The graph below shows the depth of water, in metres, in a tank after t seconds.

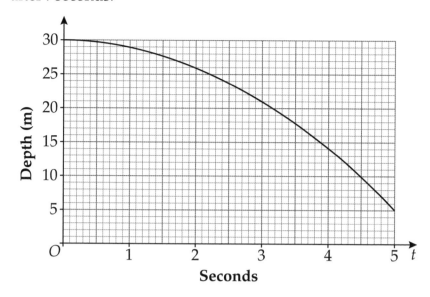

(a) Work out the average rate of the change in depth from $t = 0$ to $t = 3$

.... m/s

(2)

(b) Work out an estimate for the rate of change of depth at $t = 3$

.... m/s

(3)

(c) Explain why your answer to part **(b)** is only an estimate.

....

(1)

(Total for Question 13 is 6 marks)

Turn to page 143 for complete worked solutions to the questions on this page.

√xy² **ALGEBRA**

Revision Guide
Page 54

Hint

For part **(a)**, find the depth at $t = 0$ and the depth at $t = 3$

Average rate of change
$$= \frac{\text{Change in depth}}{\text{Time}}$$

Hint

For part **(b)**, draw a **tangent** to the curve at the point where $t = 3$. The gradient of the tangent is an estimate for the rate of change.

Hint

The gradient is negative, but you can give your rate of change as a positive or a negative number.

Explore

The graph gets steeper as t increases, so the rate of change is **increasing**.

Revision Guide
Page 38

 **ALGEBRA**

Hint

To solve a **quadratic inequality**:
- rearrange so one side is 0
- factorise the other side
- sketch a graph
- write down the values that satisfy the inequality.

Hint

The inequality is < 0 so you are interested in the values on your graph where the curve is **below** the x-axis.

Hint

Give your answer as a pair of inequalities, or using set notation.

14 Solve $x^2 < 6x - 8$

Scan this QR code for a video of this question being solved!

....
(Total for Question 14 is 3 marks)

Turn to page 143 for complete worked solutions to the questions on this page.

15 The incomplete table shows information about the times, in minutes, that runners took to complete a race.

Time (t minutes)	$30 \leqslant t < 35$	$35 \leqslant t < 40$	$40 \leqslant t < 50$	$50 \leqslant t < 60$	$60 \leqslant t < 80$
Number of runners	12	20		12	16

(a) Use the histogram to calculate the number of runners who took between 40 and 50 minutes to complete the race.

....

(2)

(b) Complete the histogram for the remaining results.

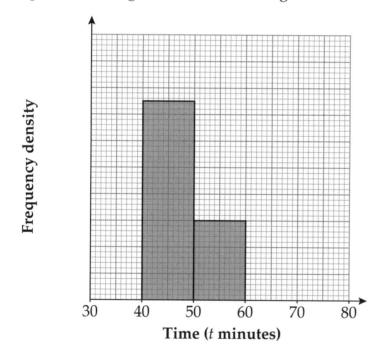

Frequency density

Time (t minutes)

(2)
(Total for Question 15 is 4 marks)

PROBABILITY & STATISTICS

REVISION GUIDE Revision Guide Page 120

LEARN IT!

On a histogram, **area** is proportional to frequency.

LEARN IT!

The vertical axis on a histogram always represents **frequency density**.

LEARN IT!

Frequency density

$= \dfrac{\text{Frequency}}{\text{Class width}}$

Problem solved!

You can often solve histogram problems by working out the scale on the vertical axis. Calculate the frequency density for the $50 \leqslant t < 60$ class, and use that bar to work out what each large grid square represents.

Turn to page 143 for complete worked solutions to the questions on this page.

77

NUMBER

 ALGEBRA

 Revision Guide
Pages 5, 21

Watch out!

This is a **non-calculator** paper, so you need to use a written method to simplify the fractions.

Hint

$$\frac{1}{3.5} = \frac{1}{\left(\frac{7}{2}\right)} = \frac{2}{7}$$

Hint

Substitute into the formula and simplify to find $\frac{1}{f}$. Then find the reciprocal to find f.

---

16 The Lens formula is used to work out the focal length.

$$\frac{1}{f} = \frac{1}{u} + \frac{1}{v}$$

where u is the distance of an object from a lens, v is the distance of the image from the lens and f is the focal length of the lens.

Work out the focal length, f, when $u = 3.5$ cm and $v = -4$ cm.

.... cm

(Total for Question 16 is 3 marks)

Turn to page 143 for complete worked solutions to the questions on this page.

17

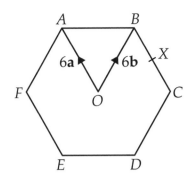

The diagram shows a regular hexagon $ABCDEF$ with centre O.

$$\overrightarrow{OA} = 6\mathbf{a} \qquad \overrightarrow{OB} = 6\mathbf{b}$$

(a) Express in terms of $\mathbf{a}$ and $\mathbf{b}$

  (i) $\overrightarrow{AB}$

                             ....

  (ii) $\overrightarrow{EF}$

                             ....
                                          (2)

X is the midpoint of BC.

(b) Express $\overrightarrow{EX}$ in terms of $\mathbf{a}$ and $\mathbf{b}$

                             ....
                                          (2)

Y is the point of AB extended, such that $AB : BY = 3 : 2$

(c) Prove that E, X and Y lie on the same straight line.

(2)

(Total for Question 17 is 6 marks)

NUMBER

ALGEBRA

Revision Guide
Pages 106, 107

Hint

To find an expression for $\overrightarrow{AB}$, trace a path from A to B. If you go backwards along a vector you **subtract** it.

Hint

The shape is a regular hexagon so $\overrightarrow{EF} = \overrightarrow{OA}$

Problem solved!

For part **(c)**, extend the line AB on the drawing, and mark point Y.

Hint

$$\overrightarrow{BY} = \frac{2}{3}\overrightarrow{AB}$$

Hint

To show that E, X and Y lie on the same straight line you need to show that any two of the vectors $\overrightarrow{EX}$, $\overrightarrow{XY}$ and $\overrightarrow{EY}$ are parallel.

Turn to page 144 for complete worked solutions to the questions on this page.

 **ALGEBRA**

 Revision Guide
Pages 17, 52

Problem solved!

The question says "Prove that…" so you need to show all your working clearly and write a conclusion.

Hint

Multiply out the brackets then simplify the expression.

Hint

To show that something is a multiple of 8 you need to show that you can write it as 8 times an integer.

 **Explore**

Test this result for some different values of n. For example, when $n = 4$:

$(2n + 3)^2 - (2n - 3)^2$
$= 11^2 - 5^2$
$= 121 - 25$
$= 96 = 8 \times 12$

18 Prove that $(2n + 3)^2 - (2n - 3)^2$ is a multiple of 8 for all positive integer values of n.

 Scan this QR code for a video of this question being solved!

(Total for Question 18 is 3 marks)

Turn to page 144 for complete worked solutions to the questions on this page.

19

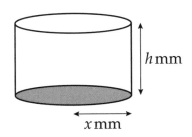

h mm

x mm

A manufacturer produces pain-relieving tablets. Each tablet is in the shape of a solid circular cylinder with base radius x mm and height h mm, as shown above in the diagram.

Given that the volume of each tablet has to be 60 mm³, show that the surface area, A mm², of a tablet is given by $A = 2\pi x^2 + \dfrac{120}{x}$

(Total for Question 19 is 4 marks)

TOTAL FOR PAPER IS 80 MARKS

 ALGEBRA

Revision Guide
Page 83

LEARN IT!

Volume of cylinder
$= \pi r^2 h$

LEARN IT!

Surface area of cylinder
$= 2\pi r^2 + 2\pi r h$

Problem solved!

Use the fact that the volume of the tablet is 60 mm³ to write h in terms of x. Then substitute this into the expression for the surface area of a cylinder.

Watch out!

Write out any formulae before you substitute.

Turn to page 144 for complete worked solutions to the questions on this page.

 Revision Guide
Page 61

Problem solved!

Work out the weight of
1 packet of crisps and
the number of calories
per gram.

Hint

You could also work
out the total number of
calories in 864 grams,
then divide by 36

Paper 2: Calculator
Time allowed: 1 hour 30 minutes

1   The total weight of 36 packets of crisps is 864 grams.

There are 505 calories in 100 grams of crisps.

Work out how many calories are there in one packet of crisps.

.... calories
(Total for Question 1 is 2 marks)

Turn to page 145 for complete worked solutions to the questions on this page.

2 A baker makes jam rolls.

The baker uses flour, butter and jam in the ratio 8 : 4 : 5 to make the jam rolls.

The table shows the cost per kilogram of some of these ingredients.

Cost per kilogram	
Flour	40p
Butter	£2.50
Jam	£1.00

The total weight of the flour, butter and jam for each jam roll is 425 g.

The baker wants to make 200 jam rolls.

He has £90 to spend on the ingredients.
Does he have enough money?
You must show your working.

RATIO AND PROPORTION

Revision Guide
Page 61

Problem solved!

Start by working out how much of each ingredient is needed to make 1 jam roll. You need to divide 425 g in the ratio 8 : 4 : 5

Hint

Multiply the weights in grams needed for 1 jam roll by 200 to find the amounts of each ingredient needed.

Watch out!

Remember to convert from grams to kg by dividing by 1000

Watch out!

Be careful with £ and pence. You should convert the cost of flour into £ before calculating.

Hint

Remember to compare your final amount with £90 and write a conclusion.

(Total for Question 2 is 5 marks)

RATIO AND PROPORTION

 Revision Guide
Page 1

Hint

For the HCF choose the lowest power of each prime factor. For the LCM choose the highest power of each prime factor.

Watch out!

Check your answer. The HCF cannot be larger than either value and the LCM cannot be smaller than either value.

Explore

$$LCM = \frac{AB}{HCF}$$

3 $A = 2^4 \times 3^2 \times 7$ $\qquad$ $B = 2^3 \times 3^4 \times 5$

A and B are numbers written as the product of their prime factors.

Find

(a) the highest common factor of A and B,

....

(2)

(b) the lowest common multiple of A and B.

....

(1)

(Total for Question 3 is 3 marks)

Turn to page 145 for complete worked solutions to the questions on this page.

4 A bank pays compound interest of 9.25% per annum.
Ravina invests £8600 for 3 years.

(a) Calculate the interest earned after 3 years.

£ ....

(3)

(b) Show that the interest gained after 3 years is 30.4% of her
original investment.

Scan this QR
code for a video
of this question
being solved!

(2)

(Total for Question 4 is 5 marks)

**RATIO AND
PROPORTION**

Revision Guide
Pages 62, 64

Hint

To find the multiplier for
a percentage increase,
add the increase to
100 then divide by 100:

$$\frac{100 + 9.25}{100} = 1.0925$$

LEARN IT!

Use this formula to
find the total amount
after n years:

$\text{starting amount} \times (\text{multiplier})^n$

Watch out!

You need to find the
interest, not the total
amount in Ravina's
account. Subtract
£8600 from the total
amount to find the
interest earned.

Turn to page 145 for complete worked solutions to the questions on this page.

GEOMETRY AND MEASURES

Revision Guide
Page 74

Watch out!

The diagram is not accurately drawn, so you have to use angle facts to determine whether the lines are parallel.

Watch out!

You don't know that the lines are parallel, so you can't use angle facts about parallel lines in your working.

Problem solved!

If you can show that two **corresponding angles** are equal then the lines are parallel:

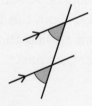

Hint

Give reasons for **every step** of your working.

---

5 The diagram shows a side view of a kitchen step ladder.

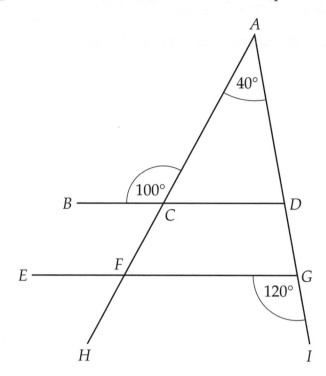

Brian says the straight lines *BCD* and *EFG* are parallel.

Is Brian correct?
You must show all your working.
Give reasons for your answer.

(Total for Question 5 is 3 marks)

Turn to page 146 for complete worked solutions to the questions on this page.

6   Prove algebraically that the recurring decimal $0.3\dot{2}$ has the value $\frac{29}{90}$

Scan this QR
code for a video
of this question
being solved!

(Total for Question 6 is 2 marks)

NUMBER

Revision Guide
Page 9

Hint

$0.3\dot{2}$ means $0.322222\ldots$

Hint

Write the number as n.
There is one recurring
digit so multiply by 10
to get $10n$. Then work
out $10n - n$ to get an
expression for $9n$.

Explore

Numbers like π and
$\sqrt{2}$ can't be written as
either a fraction or a
recurring decimal. They
are called **irrational
numbers**.

Turn to page 146 for complete worked solutions to the questions on this page.

87

GEOMETRY AND MEASURES

Revision Guide
Pages 77, 78

LEARN IT!

Remember
SOH CAH TOA:

$\sin x = \dfrac{\text{opp}}{\text{hyp}}$

$\cos x = \dfrac{\text{adj}}{\text{hyp}}$

$\tan x = \dfrac{\text{opp}}{\text{adj}}$

Hint

For part **(a)**, you know **hyp** and want the **opp** so use sin.

Hint

You need to use the $\boxed{\tan^{-1}}$ function on your calculator to find the angle in part **(b)**.

Watch out!

The height of the triangle for part **(b)** is the height of the building **minus** your answer to part **(a)**.

Hint

Write down at least 4 decimal places from your calculator display before rounding your answer.

7   PQR is the side of a vertical building.
    AB is a ramp.
    AP is horizontal ground.
    BQ is a horizontal path.

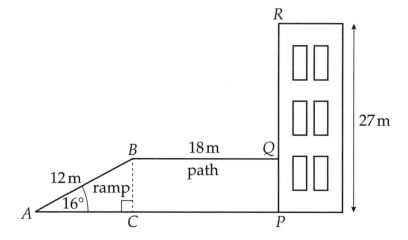

The building has a height of 27 m.
The ramp AB is at an angle of 16° to the horizontal ground.
The ramp has a length of 12 m. The path has a length of 18 m.

(a) Work out the height of the ramp.
     Give your answer correct to 3 significant figures.

..................................m
(2)

(b) Show that the angle of elevation of the top of the building, R, from the top of the ramp, B, is 52.8° correct to 3 significant figures.

(3)
(Total for Question 7 is 5 marks)

Turn to page 146 for complete worked solutions to the questions on this page.

8   A curve has an equation $y = (x - 4)(x - 10)$.

   (a) Write down the coordinates of the point where the curve cuts
        the y-axis.

$\sqrt{xy^2}$ **ALGEBRA**

Revision Guide
Pages 28, 43

Hint

Substitute $x = 0$ into
the equation to find the
y-intercept.

(...................... , ...)
                                              (1)

The curve has a turning point at A.

   (b) Work out the coordinates of A.

LEARN IT!

The quadratic
graph with equation
$y = (x - a)(x - b)$
crosses the x-axis at
$(a, 0)$ and $(b, 0)$. The
graph is **symmetrical**
so the turning point is
halfway between these
points.

Watch out!

You need to find the
x-coordinate **and the
y-coordinate** of the
turning point.

Problem solved!

Scan this QR
code for a video
of this question
being solved!

(...................... , ...)
                                              (3)
(Total for Question 8 is 4 marks)

Find the x-coordinate
of the turning point
then substitute this
into the equation to
find the y-coordinate.

Turn to page 146 for complete worked solutions to the questions on this page.

 **ALGEBRA**

 Revision Guide
Page 21

Watch out!

This looks like a speed question, but it's actually a **formulae** question.

Hint

Substitute the given values for *v* and *a* into the formula to work out Tina's stopping distance.

Hint

The units can give you a clue about which values to substitute for which variables: *a* is measured in m/s^2, *v* is measured in m/s and *s* is measured in m.

Problem solved!

Make sure you **answer the question**. Compare Tina's stopping distance with 25 m and write a short conclusion.

---

9   Tina is driving her car at a speed of 13.3 m/s.
She is at a distance of 25 m from a stop sign.

The greatest stopping distance, *s* metres, of Tina's car can be found using the formula:

$$s = \frac{v^2}{2a}$$

where

   v m/s is the speed of the car,
   a m/s^2 is the deceleration of the car when braking.

The deceleration of Tina's car when braking is 3.86 m/s^2.

Will Tina be able to stop the car before the stop sign?

(Total for Question 9 is 3 marks)

Turn to page 147 for complete worked solutions to the questions on this page.

10 The diagram shows a tank of water.
Water is leaking out of the bottom of the tank.

The graph shows the amount of water, g gallons, in the tank at time, t hours.

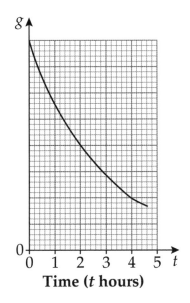

There are 8000 gallons of water in a full tank.

The number of gallons of water in the tank at time, t hours, is modelled by the formula:

$$g = ka^{-t}$$

where k and a are positive constants.

Work out the value of k and the value of a.

$k =$ ....

$a =$ ....

(Total for Question 10 is 4 marks)

Turn to page 147 for complete worked solutions to the questions on this page.

 ALGEBRA

 RATIO AND PROPORTION

Revision Guide
Pages 53, 64

Hint

When $t = 0$, $g = 8000$. Substitute these values into $g = ka^{-t}$ to find the value of k

Hint

Use the graph with $t = 2$ to find the corresponding value of g. Use these values and your value of k to find the value of a.

Watch out!

You can read a third value off the graph (such as $t = 4$) and use this to **check** your answer.

Watch out!

The power is $-t$. Be careful with the negative sign.

Explore

This is an example of **exponential decay**. The horizontal axis is an asymptote to this graph. The amount of water will get closer and closer to zero but will never reach it.

PROBABILITY & STATISTICS

Revision Guide Page 119

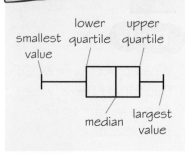

LEARN IT!

smallest value | lower quartile | upper quartile | median | largest value

Hint

Use a ruler and sharp pencil to draw your box plot in part **(c)**.

Watch out!

When comparing distributions, your comments must be **in the context** of the question. Don't just compare numbers – you need to talk specifically about **cars** and **miles per gallon**.

Hint

Compare one average (the **median**) and one measure of spread (the **range** or the **interquartile range**).

---

11 The box plot shows information, from 2006, about the distribution of the average miles per gallon (mpg) from a random sample of cars.

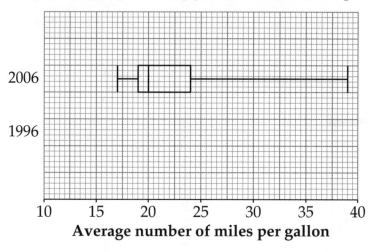

(a) Write down the median for the sample of cars in 2006.

.....................................mpg
(1)

(b) Work out the interquartile range for the sample of cars in 2006

.....................................mpg
(2)

The table below shows information, from 1996, about the distribution of the average number of miles per gallon (mpg) from a random sample of cars.

	Smallest	Lower quartile	Median	Upper quartile	Largest
Average number of miles per gallon	15	17	18	22	30

(c) On the grid above, draw a box plot to show the information in the table.

(3)

(d) Compare these two distributions.

....

....

....

....

(2)

(Total for Question 11 is 8 marks)

Turn to page 147 for complete worked solutions to the questions on this page.

12 (a) $x^6 = 1000$

Find the value of x.
Give your answer correct to 3 significant figures.

$x = $ ....
(1)

(b) $y^{\frac{1}{2}} = 1000$

Find the value of y.

$y = $ ....
(1)

(c) $z^{\frac{2}{3}} = 256$

Find the value of z.

$z = $ ....
(1)
(Total for Question 12 is 3 marks)

 ALGEBRA

Revision Guide
Page 16

Hint

Write down at least 4 decimal places from your calculator display before rounding to 3 significant figures.

Hint

The inverse operation of 'raised to the power 6' is 'sixth root' or 'raised to the power $\frac{1}{6}$'.

Hint

$\left(y^{\frac{1}{2}}\right)^2 = y$

Explore

Use the $\sqrt[\square]{\square}$ function on your calculator to work out $\sqrt[6]{1000}$. Then use the $x^{\square}$ function and the fraction key to work out $1000^{\frac{1}{6}}$. Check that you get the **same answer**.

Turn to page 147 for complete worked solutions to the questions on this page.

% **RATIO AND PROPORTION**

 Revision Guide
Pages 69, 70

LEARN IT!

If t is directly proportional to the square root of L, the proportionality formula is:

$t = k\sqrt{L}$

Hint

Use $t = 2$ and $L = 100$ to find the value of k. Then use your formula to find t when $L = 64$

Explore

The exact formula for the period of a pendulum is $t = 2\pi\sqrt{\dfrac{L}{g}}$, where L is measured in **metres** and g is the acceleration due to gravity ($9.8\,\text{m/s}^2$ on Earth). Use this formula to check your answer. Remember to convert $64\,\text{cm}$ to m first.

13 The time, t seconds, it takes a pendulum to swing from its start position and back to its start position is directly proportional to the square root of its length, $L\,\text{cm}$.

A pendulum with a length of 100 cm takes 2 seconds to swing from its start position and back to its start position.

A different pendulum has a length of 64 cm.

Will it take longer for this pendulum to swing from its start position and back to its start position compared to the pendulum with a length of 100 cm?

You must show all your working.

....

(Total for Question 13 is 4 marks)

Turn to page 148 for complete worked solutions to the questions on this page.

14 (a) Show that:

$$4 \qquad 7 \qquad 12 \qquad 19 \qquad 28$$

is a quadratic sequence.

 √xy² ALGEBRA

Revision Guide
Page 24

LEARN IT!

If a sequence is **quadratic**, then the **second difference** will be constant.

(2)

LEARN IT!

(b) Hence write down an expression for the nth term of this sequence.

If the second difference is k, then the coefficient of n^2 in the nth term will be $\frac{1}{2}k$

Hint

Compare each term of the sequence with $\frac{1}{2}kn^2$ to spot the general term.

Watch out!

Check your answer by substituting some values of n. For example, when $n = 4$ the nth term should be 19

....

(2)

(Total for Question 14 is 4 marks)

Turn to page 148 for complete worked solutions to the questions on this page.

 Revision Guide
Pages 123, 127

Problem solved!

Start by working out P(H) and P(T). These probabilities must add up to 1 and are divided in the ratio 2 : 1

Hint

You could draw a tree diagram to represent the possible outcomes when the coin is flipped twice.

Hint

You need to consider two different outcomes: Heads then Tails and Tails then Heads.

Hint

The two flips are **independent events** so you can multiply the probabilities:

P(H then T) = P(H) × P(T)

---

15 Jim flips a biased coin.
The probability that it will land on heads is twice the probability that it will land on tails.

Jim flips the coin twice.
Find the probability that it will land once on heads and once on tails.

....
(Total for Question 15 is 4 marks)

Turn to page 148 for complete worked solutions to the questions on this page.

16 Here are two plant pots.

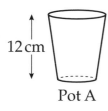

12 cm

Pot A

18 cm

Pot B

Pot A and pot B are mathematically similar.

Pot A has a height of 12 cm.
Pot B has a height of 18 cm.

Pot A has a volume of 1000 cm³.

Work out the volume of pot B.

Scan this QR code for a video of this question being solved!

.... cm³

(Total for Question 16 is 3 marks)

 GEOMETRY AND MEASURES

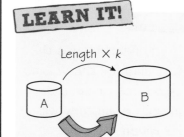 Revision Guide
Page 98

LEARN IT!

Length × k

A   B

Volume × k³

Hint

Find the length multiplier, k, then multiply the volume of A by k³.

Explore

Volume increases much more quickly than length. If you double all the dimensions of a box, it can hold 8 times as much.

Turn to page 148 for complete worked solutions to the questions on this page.

Hint

If you need to use the formulae for the volume of a cone or sphere in your exam they will be **given with the question.**

Problem solved!

Use the volume of the hemisphere to find the value of *r*. Then use your value of *r* to find the volume of the cone.

Watch out!

You are given the volume of the **hemisphere**. You need to double it to find the volume of a sphere of radius *r*.

Watch out!

Don't use rounded values in your calculations. Use at least 4 decimal places for your value of *r*, or use the ANS button on your calculator to enter the exact value.

17 The diagram shows a hemisphere on top of a cone.

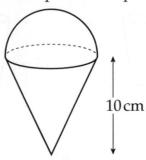

10 cm

| Volume of cone $= \frac{1}{3}\pi r^2 h$ |
| Volume of sphere $= \frac{4}{3}\pi r^3$ |

The radius of the hemisphere is equal to the radius of the cone.
The height of the cone is 10 cm.
The volume of the hemisphere is 400 cm³.

Is the volume of the cone less than the volume of the hemisphere?
You must show all your working.

(Total for Question 17 is 4 marks)

Turn to page 149 for complete worked solutions to the questions on this page.

18 The diagram shows a windscreen wiper on a car.
It also shows the area of the windscreeen the wiper cleans.

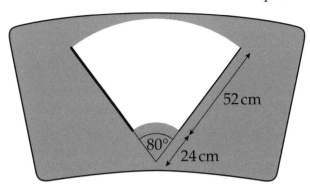

(a) Work out the area of the windscreen the wiper cleans.

......................................cm²
(3)

The area of the whole screen is 2 m².

(b) What percentage of the whole screen is cleaned by the wiper?

......................................%
(2)
(Total for Question 18 is 5 marks)

Turn to page 149 for complete worked solutions to the questions on this page.

RATIO AND PROPORTION

GEOMETRY AND MEASURES

 Revision Guide
Pages 59, 81, 84

LEARN IT!

The area of a sector of a circle with radius r and angle x is
$$\frac{x}{360°} \times \pi r^2$$

Watch out!

The area cleaned by the wiper is the **difference** of two sectors. The larger sector will have radius 52 + 24 = 76 cm.

Hint

Check that your answer makes sense. You are told in part **(b)** that the area of the whole screen is 2 m², so the area cleaned should be less than that.

Watch out!

Compare quantities in the same units.
$2 \, m^2 = 2 \times 100^2$
$\qquad = 20\,000 \, cm^2$

√xy² **ALGEBRA**

Revision Guide
Pages 27, 36,
104

LEARN IT!

The tangent to a
circle at a point is
perpendicular to the
radius at that point.

Problem solved!

Draw a sketch. It will
help you see what is
going on and check
that your answer makes
sense.

Hint

Find the gradient of
the line segment CA.
The tangent will be
perpendicular to this
line segment.

Explore

The circle will have
equation
$(x - 7)^2 + (y - 2)^2 = 10$

19   A circle has centre $C(7, 2)$.

The point $A(10, 1)$ lies on the circle.

Find the equation of the tangent at the point A.

Scan this QR
code for a video
of this question
being solved!

(Total for Question 19 is 4 marks)

Turn to page 149 for complete worked solutions to the questions on this page.

20 Solve

$$2x + y = 2$$
$$x^2 + y^2 = 1$$

√xy² **ALGEBRA**

Revision Guide
Page 35

Hint

Label your equations (I) and (2). It will make it easier to keep track of your working.

Hint

To eliminate y you need to write the first equation in the form $y = \ldots$, then substitute this expression for y into the second equation.

Watch out!

When you substitute for y^2, you need to square the **whole expression**.

Explore

Simultaneous equations involving x^2 or y^2 will usually have **two sets** of solutions. Each set is an x-value and a y-value. So you need to find four values and pair them up correctly.

(Total for Question 20 is 5 marks)

TOTAL FOR PAPER IS 80 MARKS

Turn to page 149 for complete worked solutions to the questions on this page.

 NUMBER

 Revision Guide
Page 8

Watch out!

For part **(b)**, your calculator might give the answer as an ordinary number. You have to write it in standard form.

Hint

If the power of ten is **positive** in a standard form number, then it is **greater than 1**. You are dividing by a number greater than 1, so the answer must be smaller than 2.5 × 10⁹. Check that the power of 10 in your answer is less than 9

Paper 3: Calculator
Time allowed: 1 hour 30 minutes

1 (a) Write 3.42×10^{-6} as an ordinary number.

....

(1)

(b) Work out $(2.5 \times 10^9) \div (5 \times 10^3)$

Give your answer in standard form.

....

(2)

(Total for Question 1 is 3 marks)

Turn to page 150 for complete worked solutions to the questions on this page.

2   In a sale the price of paving slabs is reduced by 70%.
Josie buys some paving slabs at the sale price of £90

What was the original price of the paving slabs?

Scan this QR
code for a video
of this question
being solved!

£ ....
(Total for Question 2 is 2 marks)

**RATIO AND
PROPORTION**

Revision Guide
Page 63

Watch out!

This is a **reverse
percentages** question.
You are given the
amount **after** the
decrease and you need
to find the **original
amount**.

Hint

Here are two possible
strategies:

I. Find the multiplier
for a 70% decrease,
then **divide** 90 by the
multiplier.

2. Divide 90 by 30 to
find 1%, then multiply by
100 to find 100%.

Hint

70% is more than half
off. Check that £90 is
less than half of your
answer.

PROBABILITY & STATISTICS

Revision Guide
Page 123

LEARN IT!

The probability of all the different outcomes of an event add up to 1

Hint

For part **(a)** add up all the probabilities, then subtract the result from 1

LEARN IT!

Expected number of outcomes =
Number of trials × probability

Watch out!

Check that your answer makes sense. It must be more than 0 and less than 40

---

3   A box contains some coloured cards.
Each card is red or blue or yellow or green.
The table shows the probability of taking a red card or a blue card or a yellow card.

Card	Probability
Red	0.3
Blue	0.35
Yellow	0.15
Green	

George takes at random a card from the box.

(a) Work out the probability that George takes a green card.

....

(2)

George replaces his card in the box.
Anish takes a card from the box and then replaces the card.
Anish does this 40 times.

(b) Work out an estimate for the number of times Anish takes a yellow card.

....

(2)

(Total for Question 3 is 4 marks)

Turn to page 150 for complete worked solutions to the questions on this page.

4   The diagram shows a right-angled triangle and a rectangle.

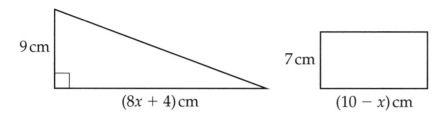

9 cm

$(8x + 4)$ cm

7 cm

$(10 - x)$ cm

The area of the triangle is twice the area of the rectangle.

Find the area of the rectangle.
Show clear algebraic working.

 ALGEBRA

Revision Guide
Page 20

LEARN IT!

Area of triangle
$= \frac{1}{2} \times$ base $\times$ height

Problem solved!

You need to use
algebra to solve
this problem. Find
expressions for the
area of each shape **in
terms of x**. Then set
the area of the triangle
equal to twice the area
of the rectangle. This
gives you an **equation**
that you can solve to
find x.

Watch out!

Read the question
carefully. After you find
x you need to work
out the **area of the
rectangle**.

Watch out!

Solutions to equations
don't have to be whole
numbers.

.....................................cm^2
(Total for Question 4 is 5 marks)

Turn to page 150 for complete worked solutions to the questions on this page.

STATISTICS

Revision Guide
Page 111

Watch out!

The modal class interval is the class interval with the highest frequency. Make sure you write down the class interval and **not** the frequency.

Hint

To find the mean, add two columns to the table. One for 'midpoint' and one for 'frequency × midpoint'.

Problem solved!

For part **(c)** think about whether the new data value will **increase** or **decrease** the mean and write a conclusion.

Explore

Your answer to part **(b)** will be an estimate because you don't know the exact data values. You are assuming that each data value is in the middle of its class interval.

5   The table shows information about the amount of money, in dollars, spent in a shop in one day by 80 people.

Money spent (x dollars)	Frequency
$0 < x \leqslant 20$	24
$20 < x \leqslant 40$	20
$40 < x \leqslant 60$	9
$60 < x \leqslant 80$	12
$80 < x \leqslant 100$	15

(a) Write down the modal class interval.

....
(2)

(b) Work out an estimate for the mean amount of money spent in that shop that day.

.... dollars
(2)

One more person spent 84 dollars.

(c) How will this affect the mean?

You must give a reason.

....

....

....

(1)
(Total for Question 5 is 5 marks)

Turn to page 151 for complete worked solutions to the questions on this page.

6 Here are three cubes.

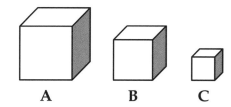

The volume of cube **B** is 20% less than the volume of cube **A**.
The volume of cube **C** is 20% less than the volume of cube **B**.
Cube **A** has a volume of 8000 cm³.

What is the volume of cube **C** as a percentage of the volume of
cube **A**?

Scan this QR
code for a video
of this question
being solved!

.....................................%

(Total for Question 6 is 3 marks)

 **RATIO AND
PROPORTION**

Revision Guide
Page 62

Hint

This looks like it might
be a question about
volume or similar
shapes. However, you
only need **percentage**
skills to answer this
question.

Hint

The multiplier for a
20% reduction is
100% − 20% = 0.8

Explore

You don't need to
know the volume of
cube **A** to answer
this question. Try the
question again, setting
the volume of cube **A**
as 12 000 cm³ instead.
What do you notice?

Turn to page 151 for complete worked solutions to the questions on this page.

PROBABILITY & STATISTICS

 Revision Guide Page 118

Hint

There were 90 people in the sample. Work out 80% of 90, then read across from this value to the curve. Read down to the horizontal axis to find the 80th percentile.

Problem solved!

If the 80th percentile is **less** than £685, then more than 80% of people earned less than this amount and the statement is true. If it is **greater** than this amount then the statement is false.

7   The cumulative frequency diagram gives information about the wages earned by a sample of 90 people in the North East of England.

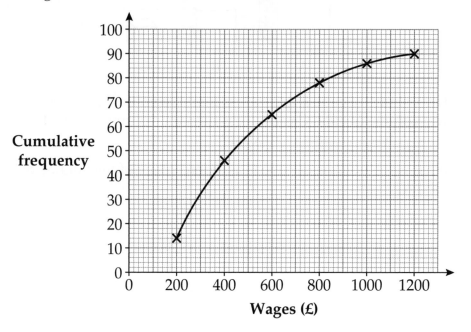

Cumulative frequency

Wages (£)

A charity claimed:

"More than 80% of the people in the North East of England earn less than £685 a week."

Is this claim correct?
You must show all your working.

(Total for Question 7 is 3 marks)

Turn to page 151 for complete worked solutions to the questions on this page.

8   Factorise $x^2 - x - 72$

....

(Total for Question 8 is 2 marks)

 ALGEBRA

Revision Guide
Page 18

Hint

Look for two numbers
that add up to −1 and
multiply to give −72

Hint

The number part is
negative, so one
number will be positive
and the other will be
negative.

Turn to page 151 for complete worked solutions to the questions on this page.

% RATIO AND PROPORTION

Revision Guide
Page 60

Problem solved!

Write the number of sweets in the jar at the start as 5x. Then there are 2x cola-flavoured sweets and 3x orange-flavoured sweets.

Hint

You can use fractions to compare quantities. If two quantities A and B are in the ratio 7 : 12 then
$$\frac{A}{B} = \frac{7}{12}$$

Watch out!

Your final answer should be the **total** number of sweets that were **originally** in the jar.

9 Simon has a jar of sweets.

The ratio of the number of cola-flavoured sweets to the number of orange-flavoured sweets in the jar is $2 : 3$

Simon eats 3 of the cola-flavoured sweets.

The ratio of the number of cola-flavoured sweets to the number of orange-flavoured sweets in the jar is now $7 : 12$

Work out the total number of sweets that were originally in the jar.

....
(Total for Question 9 is 4 marks)

Turn to page 152 for complete worked solutions to the questions on this page.

10 The diagram shows two circular cakes in a box.

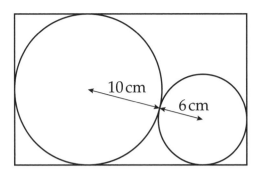

The large cake has a radius of 10 cm.
The small cake has a radius of 6 cm.

Work out the area of the base of the box.

.....................................cm²

(Total for Question 10 is 4 marks)

**GEOMETRY
AND MEASURES**

 Revision Guide
Page 76

Problem solved!

You will need to use
Pythagoras' theorem
to work out the length
of the box. Draw this
triangle on the diagram:

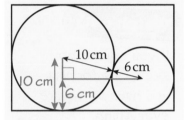

Hint

The width of the box
is the same as the
diameter of the large
circle.

Watch out!

Make sure your final
answer is the **area** of
the base of the box.

√xy² ALGEBRA

 Revision Guide
Page 46

Hint

You need to get *l* on its own on one side of the formula. Start by dividing both sides by 2π

Watch out!

If you square both sides of a formula you have to square **everything**.

Hint

Write your final answer in the form *l* = ...

11 Make *l* the subject of the formula $T = 2\pi\sqrt{\dfrac{l}{g}}$

Scan this QR code for a video of this question being solved!

....
(Total for Question 11 is 3 marks)

Turn to page 152 for complete worked solutions to the questions on this page.

12 Harry wants to buy some flooring.

He visits a warehouse.

The warehouse stocks 3 different types of flooring.

It stocks wood flooring, vinyl flooring and tile flooring.

At the ware house there are

> 30 different wood floorings,
> 24 different vinyl floorings
> and   18 different tile floorings.

Harry chooses one of each of two different types of flooring.

Work out the different number of combinations he can choose.

....
(Total for Question 12 is 3 marks)

¹₂³ NUMBER

Revision Guide
Page 13

LEARN IT!

You can work out
the total number
of combinations by
multiplying the number
of choices for each
option.

Problem solved!

You need to consider
three different
situations: wood and
vinyl, wood and tile,
vinyl and tile.

Hint

If Harry chooses wood
and tile, the number of
possible combinations
is 30 × 18

Turn to page 152 for complete worked solutions to the questions on this page.

% RATIO AND PROPORTION

 Revision Guide
Pages 68, 70

Hint

'y is inversely proportional to x' means the same as 'y is proportional to $\frac{1}{x}$'

LEARN IT!

You need to be able to recognize graphs of the following relationships:

$y \propto x$ $y \propto x^3$

$y \propto \frac{1}{x}$ $y \propto \sqrt{x}$

$y \propto x^2$ $y \propto \frac{1}{x^2}$

Watch out!

You should use a ruler and a sharp pencil when sketching graphs.

---

13 Sketch a graph of the following proportionality statements.

(a) y is inversely proportional to x.

(1)

(b) y is proportional to the square of x.

(1)
(Total for Question 13 is 2 marks)

Turn to page 153 for complete worked solutions to the questions on this page.

14

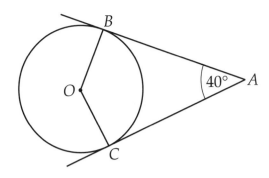

B and C are points on the circumference of a circle, centre O.
AB and AC are tangents to the circle.
Angle BAC = 40°

Find the size of angle BCO.

GEOMETRY
AND MEASURES

 Revision Guide
Pages 74, 104

LEARN IT!

The angle between a
tangent and a **radius**
is 90°

Hint

Write unknown angles
on the diagram as you
work them out.

LEARN IT!

Tangents to a circle
that meet at an
external point are
equal. So BA = CA
and triangle ABC is
isosceles.

Hint

Don't measure any
angles, but check that
your answer looks
about right.

.... °

(Total for Question 14 is 3 marks)

GEOMETRY
AND MEASURES

Revision Guide
Page 97

Hint

Triangle *ABC* is **similar** to triangle *ADE*.

Hint

Similar shapes have sides in the **same ratio**.

Watch out!

Make sure you compare the whole side length of the large triangle. For the left-hand side you need to compare 4 cm with 4 + 6 = 10 cm.

Explore

You can show the triangles are similar using the AAA property. Angle *DAE* is common to both triangles, and because *DE* is parallel to *BC*, Angle *ADE* = Angle *ABC* and Angle *AED* = Angle *ACB*.

---

15 *ABC* is a triangle.

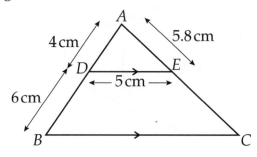

D is a point on *AB* and *E* is a point on *AC*.
DE is parallel to *BC*.
$AD = 4$ cm, $DB = 6$ cm, $DE = 5$ cm, $AE = 5.8$ cm

Calculate the perimeter of the trapezium *DBCE*.

.... cm
(Total for Question 15 is 4 marks)

Turn to page 153 for complete worked solutions to the questions on this page.

16 The average fuel consumption (c) of a car, in kilometres per litre, is given by the formula:

$$c = \frac{d}{f}$$

where d is the distance travelled in kilometres and f is the fuel used in litres.

$d = 190$ correct to 3 significant figures
$f = 25.7$ correct to 1 decimal place

By considering bounds, work out the value of c to a suitable degree of accuracy.
You must show all of your working and give a reason for your final answer.

 **NUMBER**

 **ALGEBRA**

 Revision Guide
Pages 10, 11, 21

Hint

This question uses **upper and lower bounds**. Start by writing out the upper and lower bound for each variable.

Watch out!

d is given to 3 significant figures. Its lower bound is 189.5 and its upper bound is 190.5

Problem solved!

To choose an appropriate degree of accuracy, find the maximum and minimum possible values for c. Choose the **most accurate** value that **both** round to.

Hint

Your final reason should say that the maximum and minimum of c both round to your answer, but **not** to a more accurate answer.

(Total for Question 16 is 5 marks)

Turn to page 153 for complete worked solutions to the questions on this page.

RATIO AND PROPORTION

GEOMETRY AND MEASURES

Revision Guide
Pages 66, 85

Hint

If you need to use the formula for the volume of a sphere it will be given with the question.

Hint

Draw the formula triangle for density at the top of your working:

Watch out!

Write a short conclusion to answer the question.

 **Explore**

The units of density give you a clue about how to work it out. The units are grams per cm^3 or g/cm^3. To find density you divide mass (g) by volume (cm^3).

---

17 The diagram shows a solid wooden sphere.

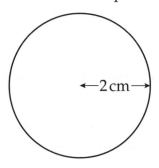

$$\text{Volume of sphere} = \tfrac{4}{3}\pi r^3$$

The radius of the sphere is 2 cm.
The mass of the sphere is 45 grams.

Wood will float on the Dead Sea only when the density of the wood is less than $1.24\,g/cm^3$.

Will this wooden sphere float on the Dead Sea?

 Scan this QR code for a video of this question being solved!

(Total for Question 17 is 4 marks)

Turn to page 154 for complete worked solutions to the questions on this page.

18 Prove that the sum of the squares of any two odd numbers is always even.

Revision Guide Page 52

√xy² **ALGEBRA**

Hint

Use $2n + 1$ to represent an odd number. If you want to represent two **different** odd numbers, use $2m + 1$ for the second one.

Hint

To show something is even write it in the form $2(…)$. To show something is odd, write it in the form $2(…) + 1$

Watch out!

Write down what you have proved as your last line of working.

Explore

Can you prove that the sum of the squares of any two odd numbers is **never** a multiple of 4?

(Total for Question 18 is 3 marks)

Turn to page 154 for complete worked solutions to the questions on this page.

GEOMETRY
AND MEASURES

Revision Guide
Pages 100, 101

LEARN IT!

Area $= \frac{1}{2}ab \sin C$

$c^2 = a^2 + b^2 - 2ab \cos C$

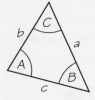

Hint

For part **(a)**, use the formula for the area of a triangle, using angle ACB and sides AC and AB.

Hint

For part **(b)** use the cosine rule with your answer to part **(a)**.

 **Explore**

You can split an isosceles triangle into two identical right-angled triangles. You *could* use this method to answer this question using only SOH CAH TOA.

19 *ABC* is an isosceles triangle.

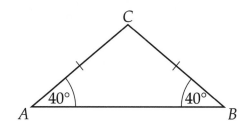

The area of this isosceles triangle is 25 cm².

(a) Work out the length of *AC*.
Give your answer correct to 3 significant figures.

.... cm

(3)

(b) Work out the length of *AB*.
Give your answer correct to 3 significant figures.

.... cm

(3)

(Total for Question 19 is 6 marks)

Turn to page 154 for complete worked solutions to the questions on this page.

20 The graphs of $y = \dfrac{4}{x}$ and $y = \dfrac{x^2}{3} - 1$ are shown below.

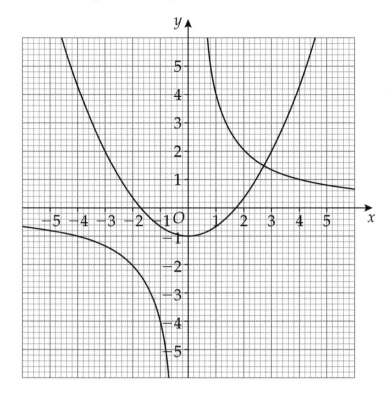

(a) Show that the x-coordinate at the point of intersection of the two graphs satisfies the equation $x^3 - 3x - 12 = 0$

(2)

(b) Show that the equation $x^3 - 3x - 12 = 0$ has a solution between $x = 2$ and $x = 3$

(2)

 ALGEBRA

Revision Guide
Page 45

Hint

For part **(a)** set the two equations equal to each other and rearrange. The solution to this equation will give you the x-coordinate of the point of intersection.

Hint

For part **(b)** you need to evaluate the function at $x = 2$ and $x = 3$. If there is a **change of sign** (from negative to positive, or vice versa) then the function has a root between 2 and 3

Turn to page 154 for complete worked solutions to the questions on this page.

Hint

At each iteration, round the value to 2 decimal places. When you find two values that round to the same number you can stop.

Watch out!

Make sure you write down the values for each step of your iteration to show your working. For this question you will need to write down values up to x_4.

(c) Use the iteration $x_{n+1} = \sqrt{\dfrac{12}{x} + 3}$ with $x_0 = 3$ to find a solution correct to two decimal places.

(3)

(Total for Question 20 is 7 marks)

Turn to page 155 for complete worked solutions to the questions on this page.

21 $y = 2x^4 - 9x^2$ and $x = \sqrt{t + 2}$ where $t > 0$

Show that t is a prime number when $y = 35$

You must show all your working.

RATIO AND PROPORTION

Revision Guide
Page 31

Hint

Substitute $y = 35$ and $x = \sqrt{t + 2}$ into the first equation.

Hint

When you are simplifying, remember that $(\sqrt{t + 2})^2 = t + 2$ and $(\sqrt{t + 2})^4 = (t + 2)^2$

Hint

You will need to solve a quadratic equation to find the value of t. Remember that $t > 0$, so your solution must be greater than 0

(Total for Question 21 is 5 marks)

TOTAL FOR PAPER IS 80 MARKS

Turn to page 155 for complete worked solutions to the questions on this page.

Paper 1: Non-calculator
Time allowed: 1 hour 30 minutes

1   Work out 3.25×0.46

Estimate: $3 \times 0.5 = 1.5$

```
      3 2 5
  ×     4 6
    1 9 5 0
  1,3 0 0 0
  1 4 9 5 0   ✓ ✓
```

4 decimal places in question so add 4 decimal

places in answer: 1.4950

1.495 ✓
....
(Total for Question 1 is 3 marks)

1

2   There are 30 children in a class.

21 of the children sing in the choir.
10 of the children play in the band.
6 of the children sing in the choir **and** play in the band.

(a) Complete the Venn diagram to show this information.

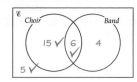

$21 - 6 = 15$
$10 - 6 = 4$
$15 + 6 + 4 = 25$
$30 - 25 = 5$

(3)

One of the children from the class is chosen at random.

(b) Work out the probability that this child plays in the band, but
does **not** sing in the choir.

$\dfrac{4}{30}$ ✓ ✓
....
(2)
(Total for Question 2 is 5 marks)

2

3   The diagram shows the area of each of three faces of a cuboid.

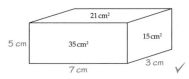

The length of each edge of the cuboid is a whole number of
centimetres.

Work out the volume of the cuboid.

$21 = 3 \times 7$
$35 = 5 \times 7$
$15 = 3 \times 5$ ✓
Volume $= 3 \times 5 \times 7 = 105$
                              ✓

105 ✓
.... cm³
(Total for Question 3 is 4 marks)

3

4   When a person exercises, their pulse rate increases.

The time it takes for their pulse rate to return to normal after
exercise is called the recovery time.

A group of people did some exercise.

The table below shows some information about their recovery times.

Recovery time (t seconds)	Cumulative frequency
$0 < t \leqslant 20$	0
$0 < t \leqslant 40$	7
$0 < t \leqslant 60$	16
$0 < t \leqslant 80$	34
$0 < t \leqslant 90$	47
$0 < t \leqslant 100$	59
$0 < t \leqslant 120$	68
$0 < t \leqslant 140$	74

(a) On the grid below, draw a cumulative frequency graph for this
information.   (2)

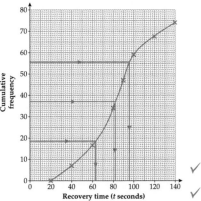

4

124

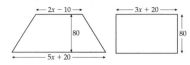
A different group of people did the same exercise.

Their recovery times had a median of 61 seconds and an interquartile range of 22 seconds.

(b) Compare the recovery times of these two groups of people.

Upper quartile (Q_3) = 96 seconds

Median (Q_2) = 82 seconds ✓

Lower quartile (Q_4) = 63 seconds
✓

Interquartile range = 96 − 63 = 33 seconds ✓

The recovery times of the first group were more spread out than those of the second group (IQR 33 s > 22 s). ✓

The recovery times of the first group were slower than the those of the second group on average (median 82 s > 61 s). ✓

Alternative acceptable answer:
Answers 1 second above or below these values.

(5)

(Total for Question 4 is 7 marks)

5

---

5 A company makes two different desks.

The top of one desk is in the shape of a trapezium.
The top of the other desk is in the shape of a rectangle.

The diagram shows the tops of the two desks.

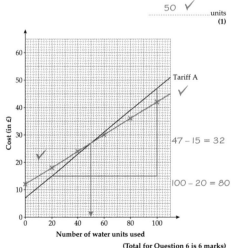

All measurements are in centimetres.

The tops of the two desks have the same area.

Work out the length, in centimetres, of the rectangular desk.
You must show all your working.

Area trapezium = $\frac{1}{2}(a + b)h$ ✓

$= \frac{1}{2}(5x + 20 + 2x − 10) × 80$

$= 40(7x + 10)$

$= 280x + 400$ ✓

Area rectangle = $80(3x + 20)$

$= 240x + 1600$ ✓

$280x + 400 = 240x + 1600$ ($−240x$)

$40x + 400 = 1600$ ($−400$)

$40x = 1200$ ($÷40$)

$x = 30$ ✓

Length of the rectangular desk

$= 3x + 20 = 3(30) + 20 = 90 + 20$

110 ✓ cm

(Total for Question 5 is 5 marks)

6

---

6 Martin's house has a meter to measure the amount of water he uses.
Martin pays on Tariff A for the number of water units he uses.

The graph on the following page can be used to find out how much he pays.

(a) (i) Find the gradient of this line.

✓

Gradient = $\frac{32}{80} = \frac{4}{10} = 0.4$

0.4 ✓

(2)

Martin reduces the amount of water he uses by 15 units.

(ii) How much money does he save?

15 × 4 = 60

15 × 0.4 = 6

£ 6 ✓

(1)

7

Instead of Tariff A, Martin could pay for his water on Tariff B.

The table shows how much Martin would pay for his water on Tariff B.

Number of water units used	0	20	40	60	80	100
Cost in £	12	18	24	30	36	42

(b) (i) On the grid, draw a line for Tariff B.

(2)

(ii) Write down the number of water units used when the cost of Tariff A is the same as the cost of Tariff B.

50 ✓ units

(1)

Tariff A ✓

47 − 15 = 32

100 − 20 = 80

(Total for Question 6 is 6 marks)

8

7 *ABCD* and *PQRS* are two rectangles.

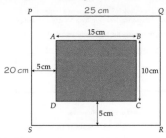

Rectangle *ABCD* is 15 cm by 10 cm.
There is a space 5 cm wide between rectangle *ABCD* and rectangle *PQRS*.

Are rectangle *ABCD* and rectangle *PQRS* mathematically similar?
You must show how you got your answer.

$PQ = 15 + 5 + 5 = 25\,cm$

$PS = 10 + 5 + 5 = 20\,cm$ ✓

$\dfrac{AB}{PQ} = \dfrac{15}{25} = 0.6$

$\dfrac{AD}{PS} = \dfrac{10}{20} = 0.5$ ✓

$0.6 \neq 0.5$ so rectangles not similar ✓

(Total for Question 7 is 3 marks)

9

---

8 (a) Write down the value of $10^{-1} \times 5^0 = \dfrac{1}{10} \times 1$ ✓

Alternative acceptable answer:
0.1

$\dfrac{1}{10}$ ✓

(2)

(b) Find the value of $27^{\frac{2}{3}}$

$\left(\sqrt[3]{27}\right)^2 = 3^2$ ✓

9 ✓

(2)

(Total for Question 8 is 4 marks)

10

---

9 Simplify fully $\dfrac{3x^2 - 6x}{x^2 + 2x - 8}$

$\dfrac{3x^2 - 6x}{x^2 + 2x - 8} = \dfrac{3x(x - 2)}{(x + 4)(x - 2)}$ ✓ ✓

$\dfrac{3x}{x + 4}$ ✓

(Total for Question 9 is 3 marks)

11

---

10 Harry travels from Appleton to Brockley at an average speed of 50 mph.
He then travels from Brockley to Cantham at an average speed of 70 mph.

Harry takes a total time of 5 hours to travel from Appleton to Cantham.
The distance from Brockley to Cantham is 210 miles.

Calculate Harry's average speed for the total distance travelled from Appleton to Cantham.

Brockley to Cantham

$T = \dfrac{D}{S} = \dfrac{210}{70} = 3$ hours ✓

Appleton to Brockley

Time $= 5 - 3 = 2$ hours

$D = S \times T = 50 \times 2 = 100$ miles ✓

Appleton to Cantham

Distance $= 100 + 210 = 310$ miles

$S = \dfrac{D}{T} = \dfrac{310}{5} = 62$ mph

✓

62 ✓ mph

(Total for Question 10 is 4 marks)

12

11

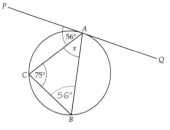

A, *B* and *C* are points on the circumference of a circle.
The straight line *PAQ* is a tangent to the circle.
Angle *PAC* = 56°
Angle *ACB* = 75°

Work out the size of the angle marked *x*.
Give reasons for each stage of your working.

Angle ABC = 56° (Alternate segment theorem) ✓

x = 180 − (75 + 56) = 49°

(Angles in a triangle add up to 180°) ✓

49 ✓ °

(Total for Question 11 is 3 marks)

---

12 The number 2.4×10^{10} is bigger than the number 6×10^{-2}

How many times bigger?
Give your answer in standard form.

$$\frac{2.4 \times 10^{10}}{6 \times 10^{-2}} = \frac{2.4}{6} \times \frac{10^{10}}{10^{-2}}$$

$$= 0.4 \times 10^{10\,-\,-2}$$

$$= 0.4 \times 10^{12} \quad ✓$$

$$= 4 \times 10^{-1} \times 10^{12}$$

4×10^{11} ✓

(Total for Question 12 is 2 marks)

13

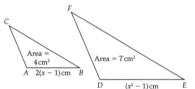

Triangles *ABC* and *DEF* are mathematically similar.

The base, *AB*, of triangle *ABC* has length $2(x-1)$ cm.
The base, *DE*, of triangle *DEF* has length (x^2-1) cm.

The area of triangle *ABC* is 4 cm².
The area of triangle *DEF* is *T* cm².

Prove that $T = x^2 + 2x + 1$

$$k = \frac{x^2 - 1}{2(x-1)} = \frac{(x+1)(x-1)}{2(x-1)} = \frac{x+1}{2} \quad ✓$$

$$k^2 = \left(\frac{x+1}{2}\right)^2 = \frac{x^2 + 2x + 1}{4} \quad ✓$$

$$T = k^2 \times 4$$

$$= 4 \times \left(\frac{x^2 + 2x + 1}{4}\right) \quad ✓$$

$$= x^2 + 2x + 1$$

So $T = x^2 + 2x + 1$ ✓

(Total for Question 13 is 4 marks)

---

14 Here is the graph of $y = \text{f}(x)$.

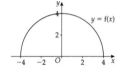

(a) Write down the coordinates of the point *P* where the graph of
$y = \text{f}(x) - 3$ meets the *y*-axis.

(0 , 1 ✓)

(1)

The graph of $\text{f}(x + 3)$ meets the negative *x*-axis at the point *A*.

(b) Work out the area of triangle *AOP*.

A = (−7, 0) ✓

$$\text{Area} = \frac{1}{2}bh = \frac{1}{2} \times 7 \times 1 = 3.5$$

✓

3.5 ✓ square units

(3)

(Total for Question 14 is 4 marks)

15 $(2x^{\frac{3}{2}}y^{-1})^n = A\,x^6\,y^B$

Work out the value of n, the value of A and the value of B.

$$\left(2x^{\frac{3}{2}}y^{-1}\right)^n = 2^n\left(x^{\frac{3}{2}}\right)^n\left(y^{-1}\right)^n$$

$$= 2^n x^{\frac{3n}{2}} y^{-n}$$

$$x^{\frac{3n}{2}} = x^6 \text{ so } \frac{3n}{2} = 6$$

$$3n = 12$$

$$n = 4$$

$2^n = A$ so $A = 2^4 = 16$

$y^{-n} = y^B$ so $B = -4$

$n = $ .... 4 ✓

$A = $ .... 16 ✓

$B = $ .... -4 ✓

(Total for Question 15 is 3 marks)

17

---

16 The resistance, R ohms, of a particular cable is inversely proportional to the square of its radius, r mm.

When the radius is 3 mm the resistance is 50 ohms.

Cable A has a radius of 5 mm.
Cable B has a radius of 10 mm.

Show that the difference in the resistance of two cables is 13.5 ohms.

$$R = \frac{k}{r^2}$$

$$50 = \frac{k}{3^2}$$

$$k = 50 \times 3^2 = 450$$

$$R = \frac{450}{r^2} \ ✓$$

$$r = 5: R = \frac{450}{5^2} = \frac{450}{25} = \frac{90}{5} = 18 \text{ ohms}$$

$$r = 10: R = \frac{450}{10^2} = \frac{450}{100} = 4.5 \text{ ohms} \ ✓$$

Difference $= 18 - 4.5 = 13.5$ ohms ✓

(Total for Question 16 is 3 marks)

18

---

17 The diagram shows an equilateral triangle ABC.

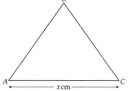

Show that the area of the triangle can be written as $\frac{x^2\sqrt{3}}{4}$

$$\text{Area} = \frac{1}{2}ab\sin C \ ✓$$

$$= \frac{1}{2}(x)(x)\sin 60°$$

$$= \frac{1}{2}x^2 \sin 60°$$

$$= \frac{1}{2}x^2\left(\frac{\sqrt{3}}{2}\right)$$

$$= \frac{x^2\sqrt{3}}{4} \ ✓$$

Alternative acceptable answer:

Height of triangle $= \sqrt{x^2 - \left(\frac{x}{2}\right)^2} = \sqrt{\frac{3}{4}x^2} = \frac{x}{2}\sqrt{3}$

Area $= \frac{1}{2}(x)\left(\frac{x}{2}\sqrt{3}\right) = \frac{x^2}{4}\sqrt{3}$

(Total for Question 17 is 2 marks)

19

18

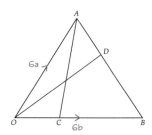

OAB is a triangle.

The point D divides the line AB in the ratio $1:2$
The point C divides the line OB in the ratio $1:2$

$\overrightarrow{OA} = 6\mathbf{a}$

$\overrightarrow{OB} = 6\mathbf{b}$

(a) Write down $\overrightarrow{AB}$ in terms of $\mathbf{a}$ and $\mathbf{b}$.

$$\overrightarrow{AB} = -\overrightarrow{OA} + \overrightarrow{OB}$$

$-6\mathbf{a} + 6\mathbf{b}$ ✓

(1)

(b) Show that:

CD is a parallel to OA **and** the length of CD is $\frac{2}{3}$ the length of OA.

$$\overrightarrow{CD} = \overrightarrow{CO} + \overrightarrow{OA} + \overrightarrow{AD}$$

$$= \frac{1}{3}(-6\mathbf{b}) + 6\mathbf{a} + \frac{1}{3}(-6\mathbf{a} + 6\mathbf{b}) \ ✓$$

$$= -2\mathbf{b} + 6\mathbf{a} - 2\mathbf{a} + 2\mathbf{b}$$

$$= 4\mathbf{a} \ ✓$$

So $\overrightarrow{CD} = \frac{2}{3}\overrightarrow{OA}$ ✓✓

So CD is parallel to OA and $\frac{2}{3}$ the length ✓

(5)

(Total for Question 18 is 6 marks)

20

19 A function is defined by $f(x) = \dfrac{x-1}{x+2}$, $x \in R$, $x \neq -2$

(a) Find $f^{-1}(x)$.

$$y = \frac{x-1}{x+2}$$
$$y(x+2) = x-1$$
$$xy + 2y = x - 1$$
$$2y + 1 = x - xy \quad \checkmark$$
$$2y + 1 = x(1-y)$$
$$x = \frac{2y+1}{1-y} \quad \checkmark$$

$$f^{-1}(x) = \underline{\quad \dfrac{2x+1}{1-x} \quad} \checkmark$$

(3)

(b) Show that $f^{-1}(x) = -2$ has no solutions.

$$\frac{2x+1}{1-x} = -2$$
$$2x + 1 = -2(1-x)$$
$$2x + 1 = 2x - 2$$
$$0x = -3 \quad \checkmark$$

Which is impossible, so no solutions $\checkmark$

(2)
(Total for Question 19 is 5 marks)

20 Given that $x = 1 + \sqrt{3}$, work out the exact value of $\dfrac{11x^2}{3-2x}$

Express your answer in the form $a + b\sqrt{3}$ where a and b are integers.

$$\frac{11(1+\sqrt{3})^2}{3-2(1+\sqrt{3})} = \frac{11(4+2\sqrt{3})}{3-2(1+\sqrt{3})} \quad \checkmark$$

$$= \frac{44 + 22\sqrt{3}}{1-2\sqrt{3}}$$

$$= \frac{(44+22\sqrt{3})(1+2\sqrt{3})}{(1-2\sqrt{3})(1+2\sqrt{3})} \quad \checkmark$$

$$= \frac{(44+22\sqrt{3})(1+2\sqrt{3})}{-11}$$

$$= \frac{44 + 110\sqrt{3} + 132}{-11} \quad \checkmark$$

$$= -16 - 10\sqrt{3}$$

$$\underline{\quad -16 - 10\sqrt{3} \quad} \checkmark$$

(Total for Question 20 is 4 marks)

TOTAL FOR PAPER IS 80 MARKS

21

22

129

Paper 2: Calculator
Time allowed: 1 hour 30 minutes

1   The Lowest Common Multiple (LCM) of three numbers is 30
    Two of the numbers are 2 and 5

    What could be the third number?

    $30 = 2 \times 3 \times 5$ ✓

    Alternative acceptable answers:
    6, 15 and 30

    ....3..... ✓
    (Total for Question 1 is 2 marks)

2   140 children will be at a school sports day.
    Lily is going to give a cup of orange drink to each of the 140 children.
    She is going to put 200 millilitres of orange drink in each cup.

    The orange drink is made from orange squash and water.
    The orange squash and water are mixed in the ratio 1 : 9 by volume.

    Orange squash is sold in bottles containing 750 millilitres.

    Work out how many bottles of orange squash Lily needs to buy.
    You must show all your working.

    $140 \times 200 = 28\,000$ ml of drink needed ✓

    $1 + 9 = 10$ parts in ratio

    $28\,000 \div 10 = 2800$ ml of squash needed ✓

    $2800 \div 750 = 3.733...$ ✓

    Lily needs to buy 4 bottles of squash. ✓
    (Total for Question 2 is 4 marks)

3   Henri and Ray buy some flowers for their mother.

    They buy:
        2 bunches of roses and 3 bunches of tulips for £10
        1 bunch of roses and 4 bunches of tulips for £9.50.

    (a) Work out the cost of one bunch of tulips.

    $2r + 3t = 10$ (1)
    $r + 4t = 9.5$ (2) ✓
    $2 \times (2):\ 2r + 8t = 19$
    $- (1):\ 2r + 3t = 10$ ✓
    $\ 5t = 9$ ✓
    $\ t = 1.8$ ✓

    £......1.80..... ✓
    (4)

    Henri is 16 years old and Ray is 2 years younger than Henri.
    They share the total cost of £19.50 in the ratio of their ages.

    (b) Work out how much Henri pays and how much Ray pays.

    $16 : 14 = 8 : 7$
    $8 + 7 = 15$
    $£19.50 \div 15 = £1.30$ ✓

    Henri: $8 \times £1.30 = £10.40$
    Ray: $7 \times £1.30 = £9.10$

    Henri £......10.40... ✓
    Ray £......9.10... ✓
    (3)
    (Total for Question 3 is 7 marks)

4   The size of each interior angle of a regular polygon with n sides is 140°

    Work out the size of each interior angle of a regular polygon with $2n$ sides.

    Exterior angle $= 180 - 140 = 40°$ ✓
    $n = 360 \div 40 = 9$ ✓

    So $2n = 18$
    Exterior angle $= 360 \div 18 = 20°$ ✓
    Interior angle $= 180 - 20 = 160°$

    ......160... ✓ °
    (Total for Question 4 is 4 marks)

23   24   25   26

130

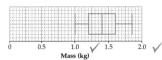
5 Zoe asked a group of 25 friends to complete two puzzles.

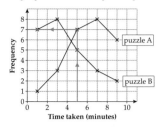

The frequency polygon shows the times taken by each of her 25 friends to complete each puzzle.

Which puzzle was harder?
Give a reason for your answer.

Only 1 person solved puzzle A in less than 2 minutes.

7 people solved puzzle B in less than 2 minutes. ✓

Puzzle A was harder. ✓

> Alternative acceptable answer:
>
> Puzzle A: 7 + 8 + 6 = 21 friends took longer than 4 minutes to solve the puzzle.
>
> Puzzle B: 5 + 3 + 2 = 12 friends took longer than 4 minutes to solve the puzzle.
>
> Puzzle A was harder.

(Total for Question 5 is 2 marks)

27

6 Here is some information about the masses, in kg, of 15 cabbages.

Smallest	1.0
Median	1.4
Upper quartile	1.6
Range	0.85
Interquartile range	0.4

On the grid draw a box plot to show this information.

Largest value = 1.0 + 0.85 = 1.85

Lower quartile = 1.6 − 0.4 = 1.2

(Total for Question 6 is 2 marks)

28

7 The nth term of sequence A is $3n - 2$
The nth term of sequence B is $10 - 2n$

Sally says there is only one number that is in both sequence A and sequence B.

Is Sally right?
You must explain your answer.

Sequence A: 1, 4, 7, 10, 13... ✓

Sequence B: 8, 6, 4, 2, 0... ✓

4 is in both sequences.

Sequence A is increasing and sequence B is decreasing so this is the only term in both sequences. Sally is correct. ✓

(Total for Question 7 is 2 marks)

29

8 x is directly proportional to y
y is inversely proportional to z

(a) Prove that z is inversely proportional to x

$x = ky$ (1)

$y = \dfrac{j}{z}$ (2) ✓

Substituting (2) into (1):

$x = k\left(\dfrac{j}{z}\right) = \dfrac{(kj)}{z}$

and $z = \dfrac{(kj)}{x}$ ✓

So z is inversely proportional to x. ✓

(3)

When $x = 40$, $z = 0.2$

(b) Work out the value of z when $x = 16$

$z = \dfrac{k}{x}$

$0.2 = \dfrac{k}{40}$

$k = 0.2 \times 40 = 8$ ✓

When $x = 16$, $z = \dfrac{8}{x} = \dfrac{8}{16} = \dfrac{1}{2}$

$z = \dfrac{1}{2}$ ✓

(2)
(Total for Question 8 is 5 marks)

30

9 On the grid below, show by shading the region defined by the inequalities

$$y > 1 \qquad y < 2x - 2 \qquad y < 6 - x \qquad x > 0$$

Mark this region with the letter R.

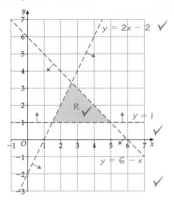

(Total for Question 9 is 4 marks)

---

10 The diagram shows a pentagon *ABCDE*.

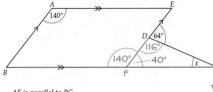

AE is parallel to *BC*.
BA is parallel to *DE*.

Angle *EDC* = 64°
Angle *BAE* = 140°

Work out the size of the angle marked *x*.
You must give reasons for your answer.

Angle *BFE* = 140°

(Opposite angles in a parallelogram are equal)

Angle *DFC* = 40°

(Angles on a straight line add up to 180°)

Angle *FDC* = 116°

(Angles on a straight line add up to 180°) ✓

x = 180° − (40° + 116°) = 24°

(Angles in a triangle add up to 180°)

✓

x = ...24....✓.... °

(Total for Question 10 is 4 marks)

---

11 Icetown makes fridges.

The probability that an Icetown fridge will have an electrical fault is 0.02
The probability that an Icetown fridge will have a mechanical fault is 0.05

(a) Complete the decision tree diagram.

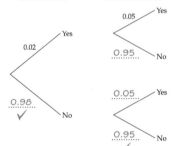

Electrical fault   **Mechanical fault**

(2)

Coolbox also makes fridges.

The probability that a Coolbox fridge will have **no** electrical fault and **no** mechanical fault is 0.93

Janet wants to buy a fridge with the least risk of any fault.

(b) Which make of fridge should Janet buy, an Icetown fridge or a Coolbox fridge?

For Icetown:

P(No fault) = 0.98 × 0.95 ✓

= 0.931 ✓

For Cooltown:

P(No fault) = 0.93

0.93 < 0.931 so Janet should by an Icetown fridge. ✓

(3)

(Total for Question 11 is 5 marks)

---

12 The functions f and g are defined as

$$f(x) = \tfrac{1}{2}x + 4$$

$$g(x) = \frac{2x}{x + 1}$$

(a) Work out f(6)

$$f(6) = \frac{1}{2}(6) + 4 = 3 + 4 = 7$$

.......7....✓....
(1)

(b) Work out fg(−3)

$$g(-3) = \frac{2(-3)}{-3 + 1} = \frac{-6}{-2} = 3$$

$$f(3) = \frac{1}{2}(3) + 4 = 1.5 + 4 = 5.5$$
✓

.......5.5....✓....
(2)

(c) g(*a*) = −2

Work out the value of *a*.

$$\frac{2a}{a + 1} = -2 \quad ✓$$

$$2a = -2(a + 1)$$

$$2a = -2a - 2$$

$$4a = -2$$

$$a = -\frac{1}{2}$$

a = ...$-\frac{1}{2}$....✓....
(2)

(d) Express the inverse function f⁻¹ in the form f⁻¹(x) = ...

$$y = \frac{1}{2}x + 4 \quad ✓$$

$$y - 4 = \frac{1}{2}x$$

$$2y - 8 = x \quad ✓$$

f⁻¹(x) = ...2x − 8....✓....
(3)

(Total for Question 12 is 8 marks)

13

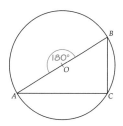

The diagram shows a circle centre O with diameter AB.

C lies on the circumference of the circle.

Prove that the angle in a semicircle is a right angle.

AB is a straight line through O so

Angle AOB = 180° ✓

Angle BCA = 180° ÷ 2 = 90° ✓

(Angle at centre of circle is twice angle at circumference) ✓

So the angle in a semicircle is a right angle. ✓

(Total for Question 13 is 4 marks)

35

14 The value of a van depreciates at the rate of 20% per year.
Gary buys a new van for £27 500
After n years the value of the van is £11 264

Find the value of n.

After 1 year: 27 500 × 0.8 = 22 000 ✓

After 2 years: 27 500 × 0.8² = 17 600

After 3 years: 27 500 × 0.8³ = 14 080

After 4 years: 27 500 × 0.8⁴ = 11 264

$n =$ ___4___ ✓

(Total for Question 14 is 2 marks)

36

15 A car accelerates from 0 metres per second to 60 metres per second in 20 seconds.
It then travels at a constant speed of 60 metres per second for 30 seconds.

The speed–time graph shows this information.

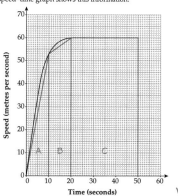

Work out an estimate for the distance the car travelled in these 50 seconds.

Area A = $\frac{1}{2}$ × 10 × 52 = 260 m

Area B = $\frac{1}{2}$ × (52 + 60) × 10 = 560 m

Area C = 30 × 60 = 1800 m

Total area = 260 + 560 + 1800 = 2620 m ✓

___2.62___ ✓ km

(Total for Question 15 is 3 marks)

37

16 The mass of substance A decreases at the rate of 20% every 6 hours.

500 grams of substance A is put into a dish.

(a) Work out the mass of substance A in the dish at the end of 12 hours.

500 × 0.8 × 0.8 = 320 ✓ ✓

___320___ ✓ grams
(3)

500 grams of a different substance B is placed in a flask.
The mass of this substance decreases at a rate of 30% every 6 hours.

At the end of 6 hours, 500 grams more of substance B is added to the flask.
At the end of a further 6 hours, another 500 grams of substance B is added to the flask.

(b) Work out the mass of substance B in the flask at the end of 18 hours.

After 6 hours

500 × 0.7 = 350

350 + 500 = 850

After 12 hours

850 × 0.7 = 595

595 + 500 = 1095

After 18 hours

1095 × 0.7 = 766.5 ✓ ✓

___766.5___ ✓ grams
(3)

(Total for Question 16 is 6 marks)

38

17 Clive wants to estimate the number of fish in a pond.

Clive catches 50 fish from the pond.
He marks each fish with a dye.
He then puts the fish back in the pond.

The next day, Clive catches 40 fish from the pond.
8 of these fish have been marked with the dye.

Work out an estimate for the number of fish in the pond.

$$N = \frac{Mn}{m} = \frac{50 \times 40}{8} = 250$$ ✓

.... 250 ✓ .... fish

(Total for Question 17 is 2 marks)

18

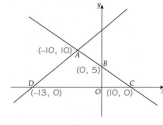

In the diagram, ABC is the line with equation $y = -\frac{1}{2}x + 5$

$AB = BC$

D is the point with coordinates $(-13, 0)$.

Find an equation of the line through A and D.

At B: $x = 0$ so $y = 5$ $B = (0, 5)$

At C: $y = 0$ so $0 = -\frac{1}{2}x + 5$

$$\frac{1}{2}x = 5$$

$$x = 10 \qquad C = (10, 0)$$ ✓

B is the midpoint of AC so $A = (-10, 10)$ ✓

Gradient of $AD = \frac{10}{3}$ ✓

Using point D: $y = mx + c$

$$0 = \frac{10}{3}(-13) + c$$ ✓

$$c = \frac{130}{3}$$

Alternative acceptable answer:
$3y = 10x + 130$

$$y = \frac{10}{3}x + \frac{130}{3}$$ ✓

(Total for Question 18 is 5 marks)

19 (a) The table and histogram show some information about the mass, in grams, of some batteries.

Use the table to complete the histogram.

Mass (m grams)	Frequency
$30 < m \leqslant 40$	4
$40 < m \leqslant 50$	6
$50 < m \leqslant 65$	15
$65 < m \leqslant 80$	9
$80 < m \leqslant 100$	4

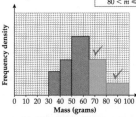

(2)

The histogram shows information about the lifetime of some batteries.

(b) Two of the batteries had a lifetime of between 1.5 and 2.5 years.
Find the total number of batteries.

1.5 to 2.5 bar = 100 small squares

50 small squares represent each battery ✓

All bars = 100 + 150 + 350 + 200 + 200
= 1000 small squares ✓

So there were 1000 ÷ 50 = 20 batteries in total ✓

(3)

(Total for Question 19 is 5 marks)

20 Rhys has a beehive.
The number of bees in the beehive is decreasing.

Rhys counts the number of bees in the hive at the start of week 5
He counts the number of bees in the hive at the start of week 7
Here are his results.

week	number bees
5	1200
7	900

Assuming that the population of bees is decreasing exponentially, how many bees were there at the start of week 2?
You must show your working.

$$900 = 1200 \times k^2$$ ✓

$$k^2 = 0.75$$

$$k = \frac{\sqrt{3}}{2}$$ ✓

If there are N bees at the start of week 2 then:

$$1200 = N \times \left(\frac{\sqrt{3}}{2}\right)^3$$

$$N = 1200 \div \left(\frac{\sqrt{3}}{2}\right)^3$$ ✓

$$= 1847.5208...$$

.... 1850 (3 s.f.) ✓ .... bees

(Total for Question 20 is 4 marks)

TOTAL FOR PAPER IS 80 MARKS

Paper 3: Calculator
Time allowed: 1 hour 30 minutes

1   The heights (in cm) of 13 girls and 13 boys were recorded.

The back-to-back stem-and-leaf diagram gives this information.

girls						boys				
		9	8	14						
		4	2	15	7	9				
8	4	④	2	0	16	2	6	8	9	
	9	5	3	0	17	⓪	3	4	6	6
					18	1	4			

KEY:
8 | 14 represents a height   15 | 7 represents a height
of 148 cm for girls   of 157 cm for boys

Compare the distribution of the heights of the girls and the distribution of the heights of the boys.

Girls' range = 179 − 148 = 31 cm
Boys' range = 184 − 157 = 27 cm
Girls' median = 164 cm
Boys' median = 170 cm ✓

Girls' heights are more spread out than the boys'
(31 cm > 27 cm). ✓

Girls were shorter on average (164 cm < 170 cm). ✓

(Total for Question 1 is 3 marks)

43

---

2   A number is increased by 25% to get 5950

What is the number?

100% + 25% = 125% ✓

$\dfrac{125\%}{100\%}$ = 1.25 = 1.25

5950 ÷ 1.25 = 4760

4760 ✓
(Total for Question 2 is 2 marks)

44

---

3   Caroline is making some table decorations. Each decoration is made from a candle and a holder.

Caroline buys some candles and some holders each in packs.

There are 30 candles in a pack of candles.
There are 18 holders in a pack of holders.

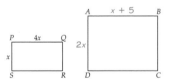

candle and holder

Caroline buys exactly the same number of candles and holders.

(a) How many packs of candles and how many packs of holders does Caroline buy?

Number of packs	1	2	3	4	5
Number of candles	30	60	⑨⓪	120	150
Number of holders	18	36	54	72	⑨⓪

✓

Alternative acceptable answers:
6 and 10, 9 and 15 and so on.

3 ✓
.... packs of candles
5 ✓
.... packs of holders
(3)

Caroline uses all her candles and all her holders.

(b) How many table decorations does Caroline make?

Alternative acceptable answers:
180, 270 and so on, to match answer to part (a).

90 ✓
.... table decorations
(1)
(Total for Question 3 is 4 marks)

45

---

4   Here are two metal plates in the shape of rectangles.

```
                          A      x + 5      B
    P    4x    Q    2x    ┌──────────────┐
    ┌──────────┐         │              │
  x │          │         │              │
    └──────────┘         └──────────────┘
    S          R    D                   C
```

In the diagram, all the measurements are in cm.

The length of AD is twice the length of PS.
The length of AB is 5 cm more than the length of PS.

Find the range of values of x for which the perimeter of the rectangle ABCD is greater than the perimeter of the rectangle PQRS.

AD = 2PS = 2x
AB = PS + 5 = x + 5

Perimeter of ABCD
2(2x) + 2(x + 5) = 4x + 2x + 10
                 = 6x + 10 ✓

Perimeter of PQRS
2(4x) + 2(x) = 10x

6x + 10 > 10x   (−6x) ✓
    10 > 4x   (÷4)
   2.5 > x   ✓

0 ⩽ x < 2.5 ✓
(Total for Question 4 is 4 marks)

46

135

5 Bill wants to compare the heights of pine trees growing in sandy soil with the heights of pine trees growing in clay soil.

The scatter diagram gives some information about the heights and the ages of some pine trees.

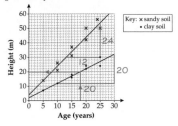

A pine tree growing in clay soil is 18 years old.

(a) Find an estimate for the height of this tree.

........................20 ✓........m
(1)

A pine tree is growing in sandy soil.

(b) Work out an estimate for how much the height of this tree increases in a year.

Gradient = $\dfrac{24}{12}$ = 2
✓

........................2 ✓........m
(2)

(c) Compare the rate of increase of the height of trees growing in clay soil with the rate of increase of the height of trees growing in sandy soil.

Gradient for clay soil = $\dfrac{20}{20}$ = 1 m / year ✓

So trees in sandy soil grow approximately twice
as fast as trees in clay soil. ✓
(2)

(Total for Question 5 is 5 marks)

47

6 A can of soup is a cylinder with a diameter 7 cm.
The can is 10 cm high.
The can is full of soup.

The soup is poured into a saucepan.
The saucepan is a cylinder with a diameter 12 cm.

Work out the depth of the soup in the saucepan.
Give your answer correct to 1 decimal place.

Volume of can = $\pi r^2 h$ ✓
$= \pi \times 3.5^2 \times 10$
$= 122.5\pi$

Soup in pan $= \pi \times 6^2 \times h$
$= 36\pi h$

$122.5\pi = 36\pi h$ ✓

$h = \dfrac{122.5\pi}{36\pi} = 3.4027...$

........................3.4 ✓........cm
(Total for Question 6 is 3 marks)

48

7 The diagram shows a chocolate bar in the shape of a triangular prism.

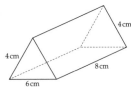

On the centimetre grid, draw a plan of the chocolate bar.

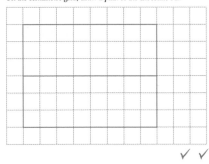

✓ ✓

(Total for Question 7 is 2 marks)

49

8 Megan is planning a game to raise money for charity.
She is going to use a fair spinner.

Megan spins the spinner twice.
The score is the sum of the numbers the spinner lands on.

(a) Complete the table to show the possible scores.

2nd spin / 1st spin	1	2	3	4
1	2	3	4	5
2	3	4	5	6
3	4	5	6	7
4	5	6	7	8

✓
(1)

Here are the rules for Megan's game:

Pay 50p to spin the spinner twice.
When the score is 7 or more, get £1.50.

(b) If this game is played 100 times should Megan expect to make a profit?

Income = 100 × 0.5 = £50 ✓

P(win) = $\dfrac{3}{16}$ ✓

Expected number of winning spins = 100 × $\dfrac{3}{16}$

$= 18.75$ ✓

Expected payout = 18.75 × 1.5 = £28.13
(nearest penny) ✓

Income > expected payout, so Megan should
expect to make a profit. ✓
(5)

(Total for Question 7 is 6 marks)

50

9   $x^2 - x - 6 \leqslant 0$

Show the solution to this inequality on the number line below.

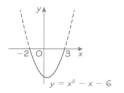

$(x - 3)(x + 2) \leq 0$ ✓

(Total for Question 9 is 4 marks)

10 The diagram shows a pond.

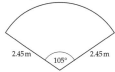

2.45 m   105°   2.45 m

The pond is in the shape of a sector of a circle.

Toby is going to put edging on the perimeter of the pond.

Edging is sold in lengths of 1.75 metres.
Each length of edging costs £3.49.

Work out the total cost of edging Toby needs to buy.

Arc length = $\dfrac{105}{360} \times 2\pi \times 2.45 = 4.4898...$ m ✓

Total perimeter = 4.4898... + 2.45 + 2.45
                = 9.3898... m ✓
9.3898... ÷ 1.75 = 5.3656... ✓
Toby needs to buy 6 lengths.
6 × £3.49 = £20.94 ✓

£ 20.94 ✓

(Total for Question 10 is 5 marks)

11 Jade makes a blackcurrant drink by mixing orange concentrate with water.

She mixes 15 cm³ of blackcurrant concentrate with 250 cm³ of water.

The density of blackcurrant concentrate is 1.20 g/cm³.
The density of water is 1.00 g/cm³.

Work out the density of Jade's blackcurrant drink.
Give your answer correct to 2 decimal places.

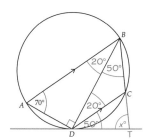

Mass of blackcurrant = 1.20 × 15 = 18 g
Mass of water = 1.00 × 250 = 250 g
Total mass = 18 + 250 = 268 g ✓

Total density = $\dfrac{268}{265}$ = 1.0113... g/cm³ ✓

1.01 ✓
g/cm³
(Total for Question 11 is 3 marks)

12

A, B, C and D are points on a circle.
AB is a diameter of the circle.
DC is parallel to AB.
Angle BAD = 70°

(a) Show that angle BDC = 20°
    You must give reasons for your working.
    Angle ADB = 90° (Angle in a semicircle = 90°) ✓
    Angle ABD = 20° (Angles in a triangle add up
                     to 180°) ✓
    Angle BDC = 20° (Alternate angles are equal) ✓
(3)

The tangent to the circle at D meets the line BC extended at T.

(b) Calculate the size of angle BTD.
    Angle BDT = 70° (Alternate segment theorem) ✓
    Angle CDT = 70° − 20° = 50°
    (Because angle BDC = 20° from part (a))
    Angle DBT = 50° (Alternate segment theorem) ✓
    So angle BTD = 180° − 70° − 50° = 60° ✓
    (Angles in triangle BDT add up to 180°) ✓
    60   °
(3)
(Total for Question 12 is 6 marks)

51

52

53

54

137

13 The area covered by the Pacific Ocean is 1.6×10^8 km².
The area covered by the Arctic Ocean is 1.4×10^7 km².

(a) Write 1.6×10^8 as an ordinary number.

160 000 000 ✓

(1)

The area covered by the Pacific Ocean is k times the area covered by the Arctic Ocean.

(b) Find, correct to the nearest integer, the value of k.

$$k = \frac{1.6 \times 10^8}{1.4 \times 10^7} = 11.4285...$$
✓

$k =$ ....11.... ✓

(2)

(Total for Question 13 is 3 marks)

14 Here is a sequence of patterns made from centimetre squares.

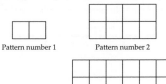

Pattern number 1   Pattern number 2

Pattern number 3

(a) Write down the number of centimetre squares used in pattern number 4.

32 ✓

(1)

(b) Find an expression, in terms of n, for the number of centimetre squares used in pattern number n.

$n \times 2n = 2n^2$
✓

$2n^2$ ✓

(2)

(c) Alex says there is a pattern in this sequence which is made from 200 centimetre squares.

Is Alex correct?
Show your working.

$2n^2 = 200 \quad (\div 2)$
$n^2 = 100 \quad (\sqrt{\ })$
$n = 10$ ✓

Pattern 10 contains 200 squares.

Yes, Alex is correct. ✓

(2)

(Total for Question 14 is 5 marks)

15 The histogram shows information about the lifetime of some electrical components.

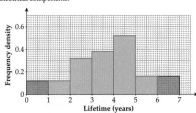

Lifetime (years)

Work out the proportion of the components with a lifetime of between 1 and 6 years.

Area between 1 and 6 years

$(0.12 \times 1) + (0.32 \times 1) + (0.38 \times 1) + (0.52 \times 1) + (0.16 \times 1) = 1.5$ ✓

Total area

$1.5 + (0.12 \times 1) + (0.16 \times 1) = 1.78$ ✓

$\frac{1.5}{1.78} = \frac{75}{89}$
✓

Alternative acceptable answers:
0.843 or
84.3% (3 s.f.)

$\frac{75}{89}$ ✓

(Total for Question 15 is 4 marks)

16 Hot drinks are served at a temperature of 70°C.

The graph shows the temperature of a hot drink as it cools in a china mug from the time it is served.

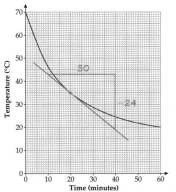

Time (minutes) ✓

Work out the rate of cooling of the drink at time 20 minutes.

$$\frac{-24}{30} = -0.8$$
✓

Alternative acceptable answers:
0.8°C per minute
Or values between −0.9 and −0.7
and values between 0.7 and 0.9

−0.8 ✓ °C per minute

(Total for Question 16 is 3 marks)

17 Expand and simplify $(x - 1)(x + 2)(x + 3)$

$(x - 1)(x + 2)(x + 3) = (x - 1)(x^2 + 5x + 6)$ ✓

$\qquad = x(x^2 + 5x + 6) - 1(x^2 + 5x + 6)$

$\qquad = x^3 + 5x^2 + 6x - x^2 - 5x - 6$ ✓

$\qquad\qquad\qquad x^3 + 4x^2 + x - 6$ ✓

(Total for Question 17 is 3 marks)

59

---

18 A car is driven through a tunnel in 89 seconds, correct to the nearest second.

The tunnel is 2460 m long, correct to the nearest 10 metres.

The average speed limit in the tunnel is 100 km/h.

Could the average speed of this car have been greater than 100 km/h? You must show your working.

	LB	UB
Time (s)	88.5	89.5
Distance (m)	2455	2465	✓

UB for speed $= \dfrac{\text{UB for distance}}{\text{LB for time}}$ ✓

$\qquad = \dfrac{2465 \div 1000}{88.5 \div 60 \div 60}$ ✓

$\qquad = 100.2711... > 100$

Average speed could have been greater than 100 km/h. ✓

(Total for Question 18 is 4 marks)

60

---

19 Here is a speed-time graph.

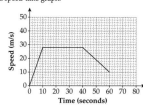

Time (seconds)

The diagram gives information about the speed of a car along a road for 60 seconds.
The speed limit for the road is 110 kilometres per hour.

(a) Does the speed of the car exceed the speed limit?

Max speed = 28 m/s ✓

28 × 3600 ÷ 1000 = 100.8 km/h ✓

The car does not exceed the speed limit. ✓

(3)

(b) Work out the acceleration of the car during the first 10 seconds.
You must write units with your answer.

Acceleration = speed ÷ time

$\qquad = 28 \div 10 = 2.8$ ✓

$\qquad\qquad\qquad 2.8$ m/s² ✓

(3)

(Total for Question 19 is 6 marks)

61

---

20 (a) Show that the equation $x^3 - 4x + 1 = 0$ can be rearranged to give

$$x = \frac{x^3}{4} + \frac{1}{4}$$

$x^3 - 4x + 1 = 0$

$\qquad x^3 + 1 = 4x$

$\qquad \dfrac{x^3}{4} + \dfrac{1}{4} = x$

As required. ✓

(1)

(b) Starting with $x_0 = 1$, use the iteration

$$x_{n+1} = \frac{x^3{}_n}{4} + \frac{1}{4}$$

to calculate the values of x_1, x_2 and x_3.

$x_0 = 1$

$x_1 = \dfrac{x_0{}^3}{4} + \dfrac{1}{4} = 0.5$ ✓

$x_2 = \dfrac{x_1{}^3}{4} + \dfrac{1}{4} = 0.28125$ ✓

$x_3 = \dfrac{x_2{}^3}{4} + \dfrac{1}{4} = 0.2555618286$ ✓

(3)

(c) Explain what the values of x_1, x_2 and x_3 represent.

Increasingly accurate estimates for a solution to $x^3 - 4x + 1 = 0$ ✓

(1)

(Total for Question 20 is 5 marks)

TOTAL FOR PAPER IS 80 MARKS

62

139

Paper 1: Non-calculator
Time allowed: 1 hour 30 minutes

1 Expand and simplify $(x + 4)(x + 6)$

$(x + 4)(x + 6) = x^2 + 6x + 4x + 24$ ✓
$\quad\quad\quad\quad\quad = x^2 + 10x + 24$

$x^2 + 10x + 24$ ✓

(Total for Question 1 is 2 marks)

---

2 Mark works for 5 days each week.

Mark can travel to work by car or train.

By car
He travels a total distance of 24 miles each day
His car travels 30 miles per gallon
Diesel costs £1.50 per litre

By train
Weekly pass costs £25.75

1 gallon = 4.5 litres

Is it more expensive if he uses his car or the train?

You must show your working.

<u>Total distance</u>
$24 \times 5 = 120$ miles ✓

<u>Diesel used</u>
$120 \div 30 = 4$ gallons
$4 \times 4.5 = 18$ litres ✓

<u>Cost of diesel</u>
$18 \times £1.50 = £27$ ✓

£27 is more than £25.75 so his car is more
expensive. ✓

(Total for Question 2 is 4 marks)

---

3 (a) Complete the table of values for $y = x^3 - 4x$

x	-3	-2	-1	0	1	2	3
y	-15	0	3	0	-3	0	15

✓ ✓
(2)

$(-3)^3 - 4(-3) = -27 + 12 = -15$
$(-2)^3 - 4(-2) = -8 + 8 = 0$
$(1)^3 - 4(1) = 1 - 4 = -3$
$(2)^3 - 4(2) = 8 - 8 = 0$

(b) On the grid, draw the graph of $y = x^3 - 4x$ from $x = -3$ to $x = 3$

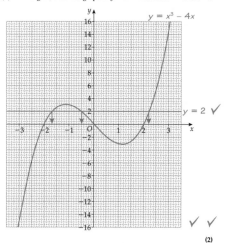

$y = x^3 - 4x$

$y = 2$ ✓

✓ ✓
(2)

(c) Use your graph to find estimates of the solutions to the equation
$x^3 - 4x = 2$

$x = -1.7, x = -0.5, x = 2.2$ ✓
(2)

(Total for Question 3 is 6 marks)

---

4 Jean makes a metal structure out of steel rope.

The diagram below shows the metal structure.

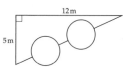

12 m
5 m

The two circles are identical.
The diameter of each circle is 2.5 metres.

Show that the total length of the steel rope used to make the metal
structure is $(a + b\pi)$ metres where a and b are integers.

Length of diagonal $= \sqrt{12^2 + 5^2} = \sqrt{169} = 13$ ✓

Wire used on diagonal $= 13 - 2.5 - 2.5 = 8$ ✓

Circumference of each circle $= \pi \times 2.5$ ✓

Total length of wire
$= 12 + 5 + 8 + 2.5\pi + 2.5\pi = 25 + 5\pi$ ✓
✓

(Total for Question 4 is 5 marks)

5 Simplify $3x^2 2y^3 \times 4xy^4 \times 2x^3y^5$

$48x^6y^{12}$ ✓ ✓

(Total for Question 5 is 2 marks)

67

6 Asha wants to buy a mobile phone.

She finds an online shop that has a sale that offers 20% off all mobile phones.

On Black Friday, the online shop reduces all sale prices by a further 30% off all mobile phones.

Asha buys a mobile phone on Black Friday.

Work out the final percentage reduction that Asha receives on the price of the mobile phone.

Multiplier for 20% reduction is × 0.8
Multiplier for 30% reduction is × 0.7 ✓
8 × 7 = 56 so 0.8 × 0.7 = 0.56 ✓

1 − 0.56 = 0.44 ✓
0.44 × 100 = 44%

44 ✓ %

(Total for Question 6 is 4 marks)

68

7 Here is a trapezium.

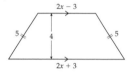

2x − 3
5   4   5
2x + 3

All the measurements are in cm.
The area of the trapezium is 18 cm².

Calculate the numerical value of the perimeter of the trapezium.

Area of trapezium $= \frac{1}{2}(a + b)h$ ✓

$= \frac{1}{2}(2x − 3 + 2x + 3) \times 4$

$= 2(4x)$

$= 8x$ ✓

$8x = 18 \qquad (\div 8)$

$x = 2.25$ ✓

Top = 2x − 3 = 4.5 − 3 = 1.5
Bottom = 2x + 3 = 4.5 + 3 = 7.5
Perimeter = 5 + 1.5 + 5 + 7.5 = 19

✓

19 ✓ cm

(Total for Question 7 is 5 marks)

69

8 Here is some information about a cricket and tennis club.

80 people belong to the club.
35 play cricket.
50 play tennis.
15 play both cricket and tennis.

(a) Draw a Venn diagram to show this information.

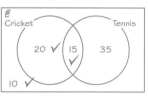

50 − 15 = 35
35 − 15 = 20
20 + 15 + 35 = 70
80 − 70 = 10

(4)

One of the people who belongs to the club is chosen at random.

(b) Work out the probability that this person does not play cricket or tennis.

Alternative acceptable answers:
$\frac{5}{40}, \frac{1}{8}$ or 0.125

$\frac{10}{80}$ ✓ ✓

(2)

(Total for Question 8 is 6 marks)

70

141

9 The distance from Caxby to Drone is 45 miles.
The distance from Drone to Elton is 20 miles.

|← 45 miles →|← 20 miles →|
Caxby   Drone   Elton

Colin drives from Caxby to Drone.
Then he drives from Drone to Elton.

Colin drives from Caxby to Drone at an average speed of 30 mph.
He drives from Drone to Elton at an average speed of 40 mph.

Work out Colin's average speed for the whole journey from Caxby
to Elton.

Caxby → Drone

$$T = \frac{D}{S} = \frac{45}{30} = 1.5 \text{ hours}$$

Drone → Elton

$$T = \frac{D}{S} = \frac{20}{40} = 0.5 \text{ hours} \checkmark$$

Whole journey

$$S = \frac{D}{T} = \frac{45 + 20}{1.5 + 0.5} = \frac{65}{2} = 32.5 \text{ mph}$$
$\checkmark$

.... 32.5 $\checkmark$ ....mph
(Total for Question 9 is 3 marks)

10

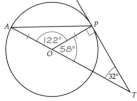

A and P are points on the circumference of a circle, centre O.
TP is a tangent to the circle.
AOT is a a straight line.
Angle $PTA = 32°$

(a) (i) State the size of angle OPT.

..........90 $\checkmark$ ......°
(1)

(ii) Give a reason for your answer.

Angle between a tangent and a radius is 90° $\checkmark$

....
....
(1)

(b) Work out the size of angle OAP.

Angle $POT = 180° - 90° - 32° = 58°$
(Angles in a triangle add up to 180°)

Angle $AOP = 180° - 58° = 122°$
(Angles on a straight line add up to 180°) $\checkmark$

Angle $OAP = (180° - 122°) \div 2 = 29°$
(Base angles in isosceles triangles are equal) $\checkmark$

..........29 $\checkmark$ ......°
(3)
(Total for Question 10 is 5 marks)

---

11 (a) Write down the value of $27^{\frac{1}{3}}$

$27^{\frac{1}{3}} = \sqrt[3]{27}$

..........3 $\checkmark$ ....
(1)

(b) Write down the value of $25^{-\frac{3}{2}}$

$$\frac{1}{25^{\frac{3}{2}}} = \frac{1}{(\sqrt{25})^3} = \frac{1}{5^3} = \frac{1}{125}$$
$\checkmark$

..........$\frac{1}{125}$ $\checkmark$ ....
(2)

(c) Simplify $\frac{(9^x)^5}{27^x}$

$$\frac{(3^2)^{5x}}{(3^3)^x} = \frac{3^{10x}}{3^{3x}} = 3^{7x}$$
$\checkmark$

..........3^{7x} $\checkmark$ ....
(2)
(Total for Question 11 is 5 marks)

---

12 (a) Express $5\sqrt{27}$ in the form $n\sqrt{3}$, where n is a positive integer.

$$5\sqrt{27} = 5\sqrt{9 \times 3}$$
$$= 5\sqrt{9} \times \sqrt{3}$$
$$= 5 \times 3 \times \sqrt{3}$$
$$= 15\sqrt{3}$$
$\checkmark$

..........$15\sqrt{3}$ $\checkmark$ ....
(2)

(b) Rationalise the denominator of $\frac{21}{\sqrt{3}}$

$$\frac{21}{\sqrt{3}} = \frac{21 \times \sqrt{3}}{\sqrt{3} \times \sqrt{3}}$$
$$= \frac{21\sqrt{3}}{3}$$
$$= 7\sqrt{3}$$
$\checkmark$

..........$7\sqrt{3}$ $\checkmark$ ....
(2)
(Total for Question 12 is 4 marks)

71   72

73   74

13 The graph below shows the depth of water, in metres, in a tank after t seconds.

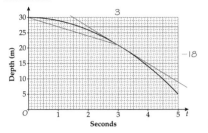

(a) Work out the average rate of the change in depth from $t = 0$ to $t = 3$

$$\frac{-9}{3} = -3 \checkmark$$

Alternative acceptable answer:

3

.... −3 $\checkmark$ m/s
(2)

(b) Work out an estimate for the rate of change of depth at $t = 3$

$$\frac{-18}{3} = -6 \checkmark$$

Alternative acceptable answers:

6

Also answers between −5.5 and −6.5 or between 5.5 and 6.5

.... −6 $\checkmark$ m/s
(3)

(c) Explain why your answer to part **(b)** is only an estimate. $\checkmark$

Because it is based on a reading off a graph.

Alternative acceptable answer:

Because the tangent isn't drawn exactly.

(1)
(Total for Question 13 is 6 marks)

75

14 Solve $x^2 < 6x - 8$

$x^2 - 6x + 8 < 0$

$(x - 4)(x - 2) < 0$ $\checkmark$

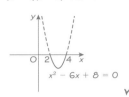

$x^2 - 6x + 8 = 0$ $\checkmark$

.... $2 < x < 4$ $\checkmark$
(Total for Question 14 is 3 marks)

76

15 The incomplete table shows information about the times, in minutes, that runners took to complete a race.

Time (t minutes)	$30 \leqslant t < 35$	$35 \leqslant t < 40$	$40 \leqslant t < 50$	$50 \leqslant t < 60$	$60 \leqslant t < 80$
Number of runners	12	20		12	16

(a) Use the histogram to calculate the number of runners who took between 40 and 50 minutes to complete the race.

Height of $50 \leqslant t < 60$ bar = 12 ÷ 10 = 1.2 $\checkmark$

So height of $40 \leqslant t < 50$ bar = 3.0

3.0 × 10 = 30

.... 30 $\checkmark$
(2)

(b) Complete the histogram for the remaining results.

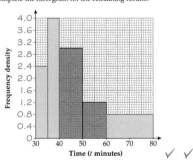

(2)
(Total for Question 15 is 4 marks)

77

16 The Lens formula is used to work out the focal length.

$$\frac{1}{f} = \frac{1}{u} + \frac{1}{v}$$

where u is the distance of an object from a lens, v is the distance of the image from the lens and f is the focal length of the lens.

Work out the focal length, f, when $u = 3.5$ cm and $v = -4$ cm.

$$\frac{1}{f} = \frac{1}{3.5} + \frac{1}{-4} \checkmark$$

$$= \frac{2}{7} - \frac{1}{4}$$

$$= \frac{8}{28} - \frac{7}{28}$$

$$= \frac{1}{28} \checkmark$$

$$f = 28$$

.... 28 $\checkmark$ cm
(Total for Question 16 is 3 marks)

78

143

17

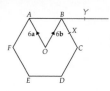

The diagram shows a regular hexagon $ABCDEF$ with centre O.

$\overrightarrow{OA} = 6\mathbf{a}$ $\qquad$ $\overrightarrow{OB} = 6\mathbf{b}$

(a) Express in terms of $\mathbf{a}$ and $\mathbf{b}$

(i) $\overrightarrow{AB}$

.... $-6a + 6b$ ✓

(ii) $\overrightarrow{EF}$

.... $6a$ ✓

(2)

X is the midpoint of BC.

(b) Express $\overrightarrow{EX}$ in terms of $\mathbf{a}$ and $\mathbf{b}$

.... ✓
.... $-3a + 12b$ ✓

(2)

Y is the point of AB extended, such that $AB : BY = 3 : 2$

(c) Prove that E, X and Y lie on the same straight line.

$\overrightarrow{XY} = 3a + \dfrac{2}{3}\overrightarrow{AB}$

$\qquad = 3a + \dfrac{2}{3}(-6a + 6b)$

$\qquad = 3a - 4a + 4b$

$\qquad = -a + 4b$

$\qquad = \dfrac{1}{3}\overrightarrow{EX}$

✓

So $\overrightarrow{XY}$ and $\overrightarrow{EX}$ are parallel. So E, X and Y are collinear. ✓

(2)

(Total for Question 17 is 6 marks)

79

---

18 Prove that $(2n + 3)^2 - (2n - 3)^2$ is a multiple of 8 for all positive integer values of n.

$(2n + 3)^2 - (2n - 3)^2$

$\qquad = (4n^2 + 12n + 9) - (4n^2 - 12n + 9)$ ✓

$\qquad = 24n$

$\qquad = 8 \times 3n$ ✓

$3n$ is an integer so $(2n + 3)^2 - (2n - 3)^2$ is a multiple of 8. ✓

(Total for Question 18 is 3 marks)

80

19

A manufacturer produces pain-relieving tablets. Each tablet is in the shape of a solid circular cylinder with base radius x mm and height h mm, as shown above in the diagram.

Given that the volume of each tablet has to be 60 mm³, show that the surface area, A mm², of a tablet is given by $A = 2\pi x^2 + \dfrac{120}{x}$

Volume $= \pi x^2 h = 60$

$\qquad h = \dfrac{60}{\pi x^2}$ ✓

Surface area $= 2\pi x^2 + 2\pi x h$ ✓

$\qquad = 2\pi x^2 + 2\pi x\left(\dfrac{60}{\pi x^2}\right)$ ✓

$\qquad = 2\pi x^2 + \dfrac{120}{x}$ ✓

(Total for Question 19 is 4 marks)

TOTAL FOR PAPER IS 80 MARKS

81

Paper 2: Calculator
Time allowed: 1 hour 30 minutes

1 The total weight of 36 packets of crisps is 864 grams.

There are 505 calories in 100 grams of crisps.

Work out how many calories are there in one packet of crisps.

$864 \div 36 = 24$ grams in 1 packet

$505 \div 100 = 5.05$ calories in 1 gram ✓

$24 \times 5.05 = 121.2$

.......121.2.....✓...... calories

(Total for Question 1 is 2 marks)

2 A baker makes jam rolls.

The baker uses flour, butter and jam in the ratio 8 : 4 : 5 to make the jam rolls.

The table shows the cost per kilogram of some of these ingredients.

Cost per kilogram	
Flour	40p
Butter	£2.50
Jam	£1.00

The total weight of the flour, butter and jam for each jam roll is 425 g.

The baker wants to make 200 jam rolls.

He has £90 to spend on the ingredients.
Does he have enough money?
You must show your working.

$8 + 4 + 5 = 17$

$425 \div 17 = 25$ ✓

<u>1 jam roll</u>

Flour: $8 \times 25 = 200$ g ✓

Butter: $4 \times 25 = 100$ g

Jam: $5 \times 25 = 125$ g

<u>200 jam rolls</u>

Flour: $0.2 \times 200 \times 0.4 = £16$ ✓

Butter: $0.1 \times 200 \times 2.50 = £50$

Jam: $0.125 \times 200 \times 1 = £25$

$16 + 50 + 25 = £91$ ✓

He does not have enough money. ✓

(Total for Question 2 is 5 marks)

3 $A = 2^4 \times 3^2 \times 7$ $B = 2^3 \times 3^4 \times 5$

A and B are numbers written as the product of their prime factors.

Find

(a) the highest common factor of A and B,

$2^3 \times 3^2 = 72$

✓

......72...✓......
(2)

(b) the lowest common multiple of A and B.

$2^4 \times 3^4 \times 5 \times 7 = 45\,360$

......45 360...✓......
(1)
(Total for Question 3 is 3 marks)

4 A bank pays compound interest of 9.25% per annum.
Ravina invests £8600 for 3 years.

(a) Calculate the interest earned after 3 years.

$8600 \times 1.0925^3 = 11\,214.06$ (2 d.p.) ✓

$11\,214.06 - 8600 = 2614.06$

✓

£......2614.06...✓......
(3)

(b) Show that the interest gained after 3 years is 30.4% of her original investment.

$\dfrac{2614.06}{8600} \times 100\% = 30.4\%$ (1 d.p.) ✓

✓

Alternative acceptable answer:
$1.0925^3 = 1.304$ so percentage increase
$= 30.4\%$

(2)
(Total for Question 4 is 5 marks)

82   83   84   85

145

5 The diagram shows a side view of a kitchen step ladder.

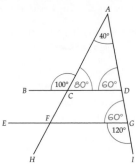

Brian says the straight lines *BCD* and *EFG* are parallel.

Is Brian correct?
You must show all your working.
Give reasons for your answer.

Angle $ACD = 180° - 100° = 80°$

(Angles on a straight line add up to 180°)

Angle $CDA = 180° - 40° - 80° = 60°$

(Angles in a triangle add up to 180°) ✓

Angle $FGD = 180° - 120° = 60°$

(Angles on a straight line add up to 180°) ✓

So corresponding angles are equal, and *BCD* and
EFG are parallel. ✓

(Total for Question 5 is 3 marks)

6 Prove algebraically that the recurring decimal $0.3\dot{2}$ has the value $\frac{29}{90}$

$10n = 3.222222...$

$- \quad n = 0.3222222...$ ✓

$9n = 2.9$

$n = \frac{2.9}{9} = \frac{29}{90}$ ✓

(Total for Question 6 is 2 marks)

7 *PQR* is the side of a vertical building.
AB is a ramp.
AP is horizontal ground.
BQ is a horizontal path.

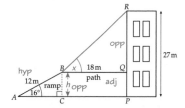

The building has a height of 27 m.
The ramp *AB* is at an angle of 16° to the horizontal ground.
The ramp has a length of 12 m. The path has a length of 18 m.

(a) Work out the height of the ramp.
Give your answer correct to 3 significant figures.

$\underline{S^O_H} \ C^A_H \ T^O_A$

$\sin 16° = \frac{h}{12}$ ✓

$h = 12 \times \sin 16° = 3.3076...$

 3.31 ✓ ....m
(2)

(b) Show that the angle of elevation of the top of the building, *R*,
from the top of the ramp, *B*, is 52.8° correct to 3 significant figures.

$S^O_H \ C^A_H \ \underline{T^O_A}$

$\tan x = \frac{27 - 3.3076...}{18}$ ✓

$= 1.3162...$

$= 52.7746...$ ✓

$x = 52.8$ ✓

(3)
(Total for Question 7 is 5 marks)

8 A curve has an equation $y = (x - 4)(x - 10)$.

(a) Write down the coordinates of the point where the curve cuts
the *y*-axis.

When $x = 0$, $y = (-4)(-10) = 40$

(............, 40.....) ✓
(1)

The curve has a turning point at A.

(b) Work out the coordinates of A.

Turning point occurs when $x = \frac{4 + 10}{2} = 7$ ✓

When $x = 7$, $y = (7 - 4)(7 - 10) = (3)(-3)$

$= -9$ ✓

 (.....7....., ...−9.....)
(3)
(Total for Question 8 is 4 marks)

9 Tina is driving her car at a speed of 13.3 m/s.
She is at a distance of 25 m from a stop sign.

The greatest stopping distance, s metres, of Tina's car can be found using the formula:

$$s = \frac{v^2}{2a}$$

where

v m/s is the speed of the car,
a m/s^2 is the deceleration of the car when braking.

The deceleration of Tina's car when braking is 3.86 m/s^2.

Will Tina be able to stop the car before the stop sign?

$$s = \frac{v^2}{2a} = \frac{13.3^2}{2 \times 3.86} = 22.9132... \quad \checkmark$$
$$\checkmark$$

22.91 m < 25 m so she will be able to stop. $\checkmark$

(Total for Question 9 is 3 marks)

10 The diagram shows a tank of water.
Water is leaking out of the bottom of the tank.

The graph shows the amount of water, g gallons, in the tank at time, t hours.

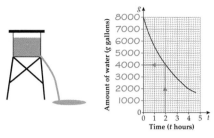

There are 8000 gallons of water in a full tank.

The number of gallons of water in the tank at time, t hours, is modelled by the formula:

$$g = ka^{-t}$$

where k and a are positive constants.

Work out the value of k and the value of a.

When $t = 0$, $g = 8000$
When $t = 2$, $g = 4000$ $\checkmark$

$8000 = ka^0$
$8000 = k$

$4000 = ka^{-2} = \dfrac{8000}{a^2}$ $\checkmark$

$a^2 = \dfrac{8000}{4000} = 2$

$a = \sqrt{2}$

$k = \underline{\quad 8000 \quad}$ $\checkmark$
$a = \underline{\quad \sqrt{2} \quad}$ $\checkmark$

(Total for Question 10 is 4 marks)

11 The box plot shows information, from 2006, about the distribution of the average miles per gallon (mpg) from a random sample of cars.

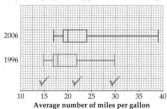

Average number of miles per gallon

(a) Write down the median for the sample of cars in 2006.

$\underline{\quad 20 \quad}$ $\checkmark$ mpg
(1)

(b) Work out the interquartile range for the sample of cars in 2006

$24 - 19 = 5$
$\checkmark$

$\underline{\quad 5 \quad}$ $\checkmark$ mpg
(2)

The table below shows information, from 1996, about the distribution of the average number of miles per gallon (mpg) from a random sample of cars.

	Smallest	Lower quartile	Median	Upper quartile	Largest
Average number of miles per gallon	15	17	18	22	30

(c) On the grid above, draw a box plot to show the information in the table.
(3)

(d) Compare these two distributions.

In 2006 cars achieved more miles per gallon (higher median). $\checkmark$

The spread of mpg was similar in both years (IQR = 5 for both years). $\checkmark$
(2)

Alternative acceptable answer:

The spread of mpg was greater in 2006 (range = 19 vs 15).

(Total for Question 11 is 8 marks)

12 (a) $x^6 = 1000$

Find the value of x.
Give your answer correct to 3 significant figures.

$x = \sqrt[6]{1000} = 3.1622...$

$x = \underline{\quad 3.16 \quad}$ $\checkmark$
(1)

(b) $y^{\frac{1}{2}} = 1000$

Find the value of y.

$y = 1000^2 = 1\,000\,000$

$y = \underline{\quad 1\,000\,000 \quad}$ $\checkmark$
(1)

(c) $z^{\frac{1}{3}} = 256$

Find the value of z.

$(\sqrt[3]{7})^2 = 256$
$\sqrt[3]{7} = \sqrt{256} = 16$
$7 = 16^3 = 4096$

$z = \underline{\quad 4096 \quad}$ $\checkmark$
(1)

(Total for Question 12 is 3 marks)

13 The time, t seconds, it takes a pendulum to swing from its start position and back to its start position is directly proportional to the square root of its length, L cm.

A pendulum with a length of 100 cm takes 2 seconds to swing from its start position and back to its start position.

A different pendulum has a length of 64 cm.

Will it take longer for this pendulum to swing from its start position and back to its start position compared to the pendulum with a length of 100 cm?

You must show all your working.

$t = k\sqrt{L}$

$2 = k\sqrt{100}$ ✓

$2 = 10k$

$k = 0.2$

$t = 0.2\sqrt{L}$ ✓

When $L = 64$:

$t = 0.2\sqrt{64}$ ✓

$= 0.2 \times 8$

$= 1.6 < 2$

No, it will take less time.   ✓

(Total for Question 13 is 4 marks)

14 (a) Show that:

$$4 \quad 7 \quad 12 \quad 19 \quad 28$$

is a quadratic sequence.

Second difference is constant so sequence is quadratic.   ✓

(2)

(b) Hence write down an expression for the nth term of this sequence.

Second difference = + 2

So $u_n = n^2 + bn + c$ ✓

n	1	2	3	4	5
u_n	4	7	12	19	28
n^2	1	4	9	16	25
$u_n - n^2$	3	3	3	3	3

So $u_n = n^2 + 3$ ✓

(2)

(Total for Question 14 is 4 marks)

15 Jim flips a biased coin.
The probability that it will land on heads is twice the probability that it will land on tails.

Jim flips the coin twice.
Find the probability that it will land once on heads and once on tails.

$P(H) = \dfrac{2}{3}$

$P(T) = \dfrac{1}{3}$ ✓

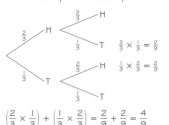

$\left(\dfrac{2}{3} \times \dfrac{1}{3}\right) + \left(\dfrac{1}{3} \times \dfrac{2}{3}\right) = \dfrac{2}{9} + \dfrac{2}{9} = \dfrac{4}{9}$ ✓

$\dfrac{4}{9}$ ✓

(Total for Question 15 is 4 marks)

16 Here are two plant pots.

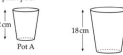

Pot A and pot B are mathematically similar.

Pot A has a height of 12 cm.
Pot B has a height of 18 cm.

Pot A has a volume of 1000 cm³.

Work out the volume of pot B.

$k = \dfrac{18}{12} = 1.5$

Volume of B = Volume of A $\times k^3$ ✓

$= 1000 \times 1.5^3$ ✓

$= 3375$

3375   ✓   cm³

(Total for Question 16 is 3 marks)

17 The diagram shows a hemisphere on top of a cone.

10 cm

Volume of cone = $\frac{1}{3}\pi r^2 h$
Volume of sphere = $\frac{4}{3}\pi r^3$

The radius of the hemisphere is equal to the radius of the cone.
The height of the cone is 10 cm.
The volume of the hemisphere is 400 cm³.

Is the volume of the cone less than the volume of the hemisphere?
You must show all your working.

$2 \times 400 = \frac{4}{3}\pi r^3$

$800 = \frac{4}{3}\pi r^3$ ✓

$600 = \pi r^3$

$r^3 = \frac{600}{\pi}$

$r = \sqrt[3]{\frac{600}{\pi}} = 5.7588...$ ✓

Volume of cone $= \frac{1}{3}\pi r^2 h$

$= \frac{1}{3} \times \pi \times 5.7588...^2 \times 10$ ✓

$= 347.2931...\,cm^3 < 400\,cm^3$

Yes, volume of cone is less than volume of
hemisphere. ✓

(Total for Question 17 is 4 marks)

---

18 The diagram shows a windscreen wiper on a car.
It also shows the area of the windscreeen the wiper cleans.

52 cm

80° 24 cm

(a) Work out the area of the windscreen the wiper cleans.

$52 + 24 = 76$

Area of large sector $= \frac{80}{360} \times \pi \times 76^2$ ✓

$= \frac{11\,552}{9}\pi$ ✓

Area of small sector $= \frac{80}{360} \times \pi \times 24^2$

$= 128\pi$

$\frac{11\,552}{9}\pi - 128\pi = 3630.2848...$

3630.3 ✓ cm²
....
(3)

The area of the whole screen is 2 m².

(b) What percentage of the whole screen is cleaned by the wiper?

$2\,m^2 = 2 \times 100^2 = 20\,000\,cm^2$

$\frac{3630.2848...}{20\,000} \times 100\% = 18.1514...$
✓

18.2 ✓ %
....
(2)
(Total for Question 18 is 5 marks)

---

19 A circle has centre $C(7, 2)$.

The point $A(10, 1)$ lies on the circle.

Find the equation of the tangent at the point A.

Gradient of $CA = -\frac{1}{3}$

Gradient of tangent $= 3$ ✓

<u>Equation of tangent</u>

$y = 3x + c$

$1 = 3(10) + c$ ✓

$c = -29$
✓

$y = 3x - 29$ ✓
....
(Total for Question 19 is 4 marks)

---

20 Solve

$2x + y = 2$ (1)
$x^2 + y^2 = 1$ (2)

From (1):

$y = 2 - 2x$

Substituting into (2):

$x^2 + (2 - 2x)^2 = 1$ ✓

$x^2 + 4 - 8x + 4x^2 = 1$

$5x^2 - 8x + 3 = 0$ ✓

$(5x - 3)(x - 1) = 0$

$x = \frac{3}{5} = 0.6$ or $x = 1$ ✓

$y = 2 - 2 \times 0.6$ $y = 2 - 2 \times 1$

$= 0.8$ $= 0$

Solutions are $x = 0.6$, $y = 0.8$ ✓
and $x = 1$, $y = 0$ ✓

(Total for Question 20 is 5 marks)

TOTAL FOR PAPER IS 80 MARKS

Paper 3: Calculator
Time allowed: 1 hour 30 minutes

1   (a) Write 3.42×10^{-6} as an ordinary number.

$0.000\,003\,42$ ✓

(1)

(b) Work out $(2.5 \times 10^{9}) \div (5 \times 10^{3}) = 500\,000$ ✓
Give your answer in standard form.

5×10^{5} ✓

(2)
(Total for Question 1 is 3 marks)

2   In a sale the price of paving slabs is reduced by 70%.
Josie buys some paving slabs at the sale price of £90
What was the original price of the paving slabs?

$100\% - 70\% = 30\%$ ✓

$\dfrac{30\%}{100\%} = 0.3$

$90 \div 0.3 = 300$

£ 300 ✓

(Total for Question 2 is 2 marks)

3   A box contains some coloured cards.
Each card is red or blue or yellow or green.
The table shows the probability of taking a red card or a blue card or a yellow card.

Card	Probability
Red	0.3
Blue	0.35
Yellow	0.15
Green	

George takes at random a card from the box.

(a) Work out the probability that George takes a green card.

$0.3 + 0.35 + 0.15 = 0.8$

$1 - 0.8 = 0.2$
✓

0.2 ✓

(2)

George replaces his card in the box.
Anish takes a card from the box and then replaces the card.
Anish does this 40 times.

(b) Work out an estimate for the number of times Anish takes a yellow card.

$0.15 \times 40 = 6$
✓

6 ✓

(2)
(Total for Question 3 is 4 marks)

4   The diagram shows a right-angled triangle and a rectangle.

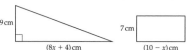

9 cm   (8x + 4) cm   7 cm   (10 − x) cm

The area of the triangle is twice the area of the rectangle.

Find the area of the rectangle.
Show clear algebraic working.

Area of triangle $= \dfrac{1}{2} \times \text{base} \times \text{height}$

$\qquad\qquad = \dfrac{1}{2}(8x + 4) \times 9$

$\qquad\qquad = (4x + 2) \times 9$

$\qquad\qquad = 36x + 18$ ✓

Area of rectangle $= \text{length} \times \text{width}$

$\qquad\qquad = (10 - x) \times 7$

$\qquad\qquad = 70 - 7x$

$36x + 18 = 2(70 - 7x)$ ✓

$36x + 18 = 140 - 14x \qquad (+ 14x)$

$50x + 18 = 140 \qquad\qquad (- 18)$

$50x = 122$

$x = 2.44$ ✓

Area of rectangle $= (10 - 2.44) \times 7$

$\qquad\qquad = 7.56 \times 7 = 52.92\,\text{cm}^2$
✓

52.92 ✓ cm²

(Total for Question 4 is 5 marks)

5 The table shows information about the amount of money, in dollars, spent in a shop in one day by 80 people.

Money spent (x dollars)	Frequency	Midpoint, x	$f \times x$
$0 < x \leq 20$	24	10	24 × 10 = 240
$20 < x \leq 40$	20	30	20 × 30 = 600
$40 < x \leq 60$	9	50	9 × 50 = 450
$60 < x \leq 80$	12	70	12 × 70 = 840
$80 < x \leq 100$	15	90	15 × 90 = 1350
Totals	80		3480

(a) Write down the modal class interval.

$0 < x \leq 20$ ✓
(2)

(b) Work out an estimate for the mean amount of money spent in that shop that day.

$\dfrac{3480}{80} = 43.5$ ✓

43.5 ✓ ....dollars
(2)

One more person spent 84 dollars.

(c) How will this affect the mean?
You must give a reason.

84 is greater than 43.5 so the mean will increase. ✓

(1)
(Total for Question 5 is 5 marks)

6 Here are three cubes.

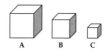

A   B   C

The volume of cube **B** is 20% less than the volume of cube **A**.
The volume of cube **C** is 20% less than the volume of cube **B**.
Cube **A** has a volume of 8000 cm³.

What is the volume of cube **C** as a percentage of the volume of cube **A**?

$100\% - 20\% = 80\% = 0.8$
Volume of B = 8000 × 0.8 = 6400 ✓
Volume of C = 6400 × 0.8 = 5120 ✓

$\dfrac{5120}{8000} \times 100\% = 64\%$

64 ✓ ....%
(Total for Question 6 is 3 marks)

7 The cumulative frequency diagram gives information about the wages earned by a sample of 90 people in the North East of England.

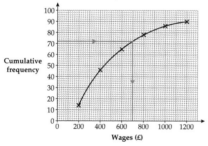

A charity claimed:

"More than 80% of the people in the North East of England earn less than £685 a week."

Is this claim correct?
You must show all your working.

80% of 90 = $\dfrac{80}{100} \times 90 = 72$ ✓

From graph 80% earn less than £700 ✓

So less than 80% earn less than £685

So the claim is not correct. ✓

(Total for Question 7 is 3 marks)

8 Factorise $x^2 - x - 72$

$8 \times -9 = -72$

$8 + -9 = -1$

✓
$(x + 8)(x - 9)$ ✓
(Total for Question 8 is 2 marks)

9 Simon has a jar of sweets.

The ratio of the number of cola-flavoured sweets to the number of orange-flavoured sweets in the jar is 2 : 3

Simon eats 3 of the cola-flavoured sweets.

The ratio of the number of cola-flavoured sweets to the number of orange-flavoured sweets in the jar is now 7 : 12

Work out the total number of sweets that were originally in the jar.

<u>At start</u>

$2x$ cola sweets and $3x$ orange sweets ✓

<u>After Simon eats 3 sweets</u>

$2x - 3$ cola sweets and $3x$ orange sweets

$$\frac{2x - 3}{3x} = \frac{7}{12}$$ ✓

$24x - 36 = 21x$

$3x = 36$

$x = 12$ ✓

$5x = 5 \times 12 = 60$

60 ✓

(Total for Question 9 is 4 marks)

---

10 The diagram shows two circular cakes in a box.

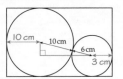

The large cake has a radius of 10 cm.
The small cake has a radius of 6 cm.

Work out the area of the base of the box.

$10 - 6 = 4\ cm$

$x = \sqrt{16^2 - 4^2} = 15.4919...$

Length of box $= 10 + 15.4919... + 3 = 28.4919...$

Area of box $= 28.4919... \times 20$ ✓

$= 569.8386...\ cm^2$

570 ✓ ....cm²

(Total for Question 10 is 4 marks)

---

11 Make l the subject of the formula $T = 2\pi\sqrt{\dfrac{l}{g}}$

$$\frac{T}{2\pi} = \sqrt{\frac{l}{g}}$$ ✓

$$\left(\frac{T}{2\pi}\right)^2 = \frac{l}{g}$$ ✓

$$g\left(\frac{T}{2\pi}\right)^2 = l$$

$$l = g\left(\frac{T}{2\pi}\right)^2$$ ✓

(Total for Question 11 is 3 marks)

---

12 Harry wants to buy some flooring.

He visits a warehouse.

The warehouse stocks 3 different types of flooring.

It stocks wood flooring, vinyl flooring and tile flooring.

At the ware house there are

        30 different wood floorings,
        24 different vinyl floorings
and   18 different tile floorings.

Harry chooses one of each of two different types of flooring.

Work out the different number of combinations he can choose.

$(30 \times 24) + (30 \times 18) + (24 \times 18) = 1692$

✓ ✓

1692 ✓

(Total for Question 12 is 3 marks)

13 Sketch a graph of the following proportionality statements.

(a) y is inversely proportional to x.

$y \propto \frac{1}{x^2}$ ✓

(1)

(b) y is proportional to the square of x.

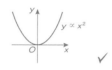

$y \propto x^2$ ✓

(1)

(Total for Question 13 is 2 marks)

14

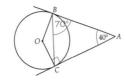

B and C are points on the circumference of a circle, centre O.
AB and AC are tangents to the circle.
Angle $BAC = 40°$

Find the size of angle BCO.

Angle ABO = Angle ACO = 90°

(Angle between a tangent and a radius = 90°) ✓

Angle ACB = Angle ABC = (180° − 40°) ÷ 2 = 70°

(Base angles of an isosceles triangle are equal)

Angle BCO = 90° − 70° = 20°
✓

20 ✓ °

(Total for Question 14 is 3 marks)

15 ABC is a triangle.

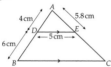

D is a point on AB and E is a point on AC.
DE is parallel to BC.
$AD = 4$ cm, $DB = 6$ cm, $DE = 5$ cm, $AE = 5.8$ cm

Calculate the perimeter of the trapezium $DBCE$.

$\frac{BC}{DE} = \frac{AB}{AD}$

$\frac{BC}{5} = \frac{10}{4}$

$BC = \frac{10 \times 5}{4} = 12.5$ cm ✓

$\frac{AC}{AE} = \frac{AB}{AD}$

$\frac{AC}{5.8} = \frac{10}{4}$

$AC = 10 \times \frac{5.8}{4} = 14.5$ cm

$EC = 14.5 − 5.8 = 8.7$ cm
✓

Perimeter $BDEC$ = 6 + 5 + 8.7 + 12.5 = 32.2 cm
✓

32.2 ✓ cm

(Total for Question 15 is 4 marks)

16 The average fuel consumption (c) of a car, in kilometres per litre, is given by the formula:

$$c = \frac{d}{f}$$

where d is the distance travelled in kilometres and f is the fuel used in litres.

$d = 190$ correct to 3 significant figures
$f = 25.7$ correct to 1 decimal place

By considering bounds, work out the value of c to a suitable degree of accuracy.
You must show all of your working and give a reason for your final answer.

	LB	UB
d	189.5	190.5
f	25.65	25.75

✓

Max for $c = \frac{190.5}{25.65} = 7.42690...$ ✓

Min for $c = \frac{189.5}{25.75} = 7.35922...$ ✓

$c = 7.4$ (1 d.p.) ✓

Max and min both round to 7.4 (1 d.p.), but round to different values to 2 d.p. ✓

(Total for Question 16 is 5 marks)

114   115

116   117

153

17 The diagram shows a solid wooden sphere.

←—2cm—→

Volume of sphere = $\frac{4}{3}\pi r^3$

The radius of the sphere is 2 cm.
The mass of the sphere is 45 grams.

Wood will float on the Dead Sea only when the density of the wood is less than 1.24 g/cm³.

Will this wooden sphere float on the Dead Sea?

$V = \frac{4}{3}\pi r^3 = \frac{4}{3}\pi \times 2^3 = 33.5103...\ cm^3$ ✓

$D = \frac{M}{V} = \frac{45}{33.5103...} = 1.3428...$ ✓

✓

1.3428... > 1.24 so the sphere will not float. ✓

(Total for Question 17 is 4 marks)

18 Prove that the sum of the squares of any two odd numbers is always even.

$(2n + 1)^2 = (2n + 1)(2n + 1)$

$= 4n^2 + 4n + 1$

$= 2(2n^2 + 2n) + 1$ ✓

So odd² = odd ✓

So odd² + odd² = odd + odd = even ✓

Alternative acceptable answer:

$(2n + 1)^2 + (2m + 1)^2$

$= 4n^2 + 4n + 4m^2 + 4m + 2$

$= 2(2n^2 + 2n + 2m^2 + 2m + 1)$

So sum of two odd numbers is even.

(Total for Question 18 is 3 marks)

19 *ABC* is an isosceles triangle.

The area of this isosceles triangle is 25 cm².

(a) Work out the length of *AC*.
Give your answer correct to 3 significant figures.

Area = $\frac{1}{2}ab\sin C$ ✓

$25 = \frac{1}{2} \times AC \times CB \times \sin 100°$

$25 = \frac{1}{2} \times AC^2 \times \sin 100°$

$AC^2 = \frac{50}{\sin 100°} = 50.7713...$ ✓

$AC = 7.1254...\ cm$

7.13 ✓
.... cm
(3)

(b) Work out the length of *AB*.
Give your answer correct to 3 significant figures.

$c^2 = a^2 + b^2 - 2ab\cos C$ ✓

$AB^2 = 7.1254...^2 + 7.1254...^2$

$- 2 \times 7.1254... \times 7.1254... \times \cos 100°$ ✓

$= 119.17535...$

$AB = 10.9167...$

10.9 ✓
.... cm
(3)
(Total for Question 19 is 6 marks)

20 The graphs of $y = \frac{4}{x}$ and $y = \frac{x^2}{3} - 1$ are shown below.

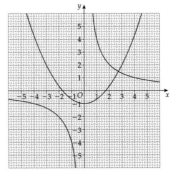

(a) Show that the x-coordinate at the point of intersection of the two graphs satisfies the equation $x^3 - 3x - 12 = 0$

$\frac{4}{x} = \frac{x^2}{3} - 1$ ✓

$4 = \frac{x^3}{3} - x$

$12 = x^3 - 3x$

$0 = x^3 - 3x - 12$ ✓

(2)

(b) Show that the equation $x^3 - 3x - 12 = 0$ has a solution between $x = 2$ and $x = 3$

When $x = 2$

$x^3 - 3x - 12 = 8 - 6 - 12 = -10$

When $x = 3$

$x^3 - 3x - 12 = 27 - 9 - 12 = 6$ ✓

Change of sign so there is a root between $x = 2$ and $x = 3$ ✓

(2)

(c) Use the iteration $x_{n+1} = \sqrt{\frac{12}{x} + 3}$ with $x_0 = 3$ to find a solution correct to two decimal places.

$x_1 = \sqrt{\frac{12}{x_0} + 3} = 2.6457... = 2.65$ (2 d.p.) ✓

$x_2 = \sqrt{\frac{12}{x_1} + 3} = 2.74509... = 2.75$ (2 d.p.)

$x_3 = \sqrt{\frac{12}{x_2} + 3} = 2.71503... = 2.72$ (2 d.p.)

$x_4 = \sqrt{\frac{12}{x_3} + 3} = 2.7239... = 2.72$ (2 d.p.) ✓

The solution is $x = 2.72$ to 2 decimal places. ✓

(3)
(Total for Question 20 is 7 marks)

---

21 $y = 2x^4 - 9x^2$ and $x = \sqrt{t + 2}$ where $t > 0$

Show that t is a prime number when $y = 35$

You must show all your working.

$35 = 2x^4 - 9x^2$

$35 = 2(\sqrt{t + 2})^4 - 9(\sqrt{t + 2})^2$ ✓

$35 = 2(t + 2)^2 - 9(t + 2)$

$35 = 2t^2 + 8t + 8 - 9t - 18$

$2t^2 - t - 45 = 0$ ✓

$(2t + 9)(t - 5) = 0$ ✓

$t = -4.5$ or $t = 5$. ✓

Since $t > 0$, $t = 5$ which is a prime number. ✓

(Total for Question 21 is 5 marks)

TOTAL FOR PAPER IS 80 MARKS

Notes

Notes

Published by Pearson Education Limited, 80 Strand, London, WC2R 0RL.

www.pearsonschoolsandfecolleges.co.uk

Copies of all official specifications for Edexcel qualifications may be found on the website:
www.edexcel.com

Text © Pearson Education Limited 2016
Edited by Andrew Briggs
Typeset and illustrated by Tech-Set Ltd, Gateshead
Produced by Out of House Publishing
Cover design by Miriam Sturdee

The rights of Jean Linsky, Navtej Marwaha and Harry Smith to be identified as authors of this work have
been asserted by them in accordance with the Copyright, Designs and Patents Act 1988.

First published 2016

19 18 17 16
10 9 8 7 6 5 4 3 2

British Library Cataloguing in Publication Data
A catalogue record for this book is available from the British Library

ISBN 9781292096315

Printed in Great Britain by Bell and Bain Ltd, Glasgow

The publisher would like to thank Edexcel for permission to use extracts from the following exam
papers: 5MB1H JUN15; 5AM1F JUN14; 5MM1F NOV14; 5AM1F JUN12; 5AM1H JUN11; 5MB3H JUN15;
5MB2H JUN13; 5MB3H NOV12; 5MM1H JUN11; 5AM1F JUN 11; AM2H JUN 11; 5MB2H NOV10;
5AM1F NOV11; 5MM2F JUN11; 5MM1F JUN 11; 5MB2H JUN 13; 5MM1H JUN 11; 5MB2H JUN 11;
5MB2H NOV15; 5MB3H NOV15; 1MAOH/2H NOV13; 1387/Paper 5 JUN03; 1MA0/2H JUN12;
5MM2H NOV15; 5AM2H NOV16; 5AM2H NOV15; 5MB3H NOV15; 4MA01F(R) MAY15; 4MA02F
MAY13; 4MA0/3H JAN 12; 4MA0/3H MAY14; 5MB3H JUN 14; 5MM1F NOV14; 5MB2H JUN15; 5MB2H
JUN14; 5MM1F JUN14; 5AM2H JUN15; IGCSE 4MA0 Paper 3HR JAN15; 5AM1H NOV14; 5MM2H
JUN15; 5AM2H JUN12; IGCSE 4MA0 Paper 3HR JUN14; 5MB1H JUN14; 5MB3H JUN12; 5AM2H
JUN14;5MB1H NOV13; 5MB2H NOV14;4MA0/1FR MAY15; 5AM2H NOV 12; 4MA0/2F JUNE 15;
1MA02F JUN 14; 5AM1H JUN 12; 1MA02F NOV 13; 5AM1H JUN12; 5MM2H JUN12; 5AM2H NOV12;
5MM1H JUN12; 5AM2H JUN13; 5MM2H NOV12; 5MB2F NOV11; 5MM2H JUN14; 5AM1F NOV11;
5MB2F NOV14;5AM1H JUN14; 5AM2H NOV13; IGCSE 4MA0 Paper 4HR JAN15; IGCSE 4MA0 Paper
4H JAN15; 5MM1H NOV11; 5AM1H NOV11; 5AM2H NOV11; 5MB2H NOV10;5MB3H NOV 14;
5MM2H JUN 11; 1MA01F NOV 13; 5MM2H NOV 13; 5AM1H JUN 14; 4MA03H JAN 13; 5MB2H JUN 14;
5MB3H NOV 13; 5AM2H JUN 14; 4MA04H(R) MAY 14; 5MM2H JUN13; 5MM2H JUN 14; Core Maths C2
AS Level MAY12; 4MA01FR MAY13.